Ongoing Efforts to Combat Illegal, Unreported, and Unregulated (IUU) Fishing

Fish, Fishing and Fisheries Series

Beyond the Blue: Smart IoT Solutions for Precision Microalgae Cultivation in Aquaculture
Polaki Suman, PhD, Chita Ranjan Sahoo, PhD and B. Rabi Prasad, PhD
ISBN: 979-8-89530-383-2
eBook ISBN: 979-8-89530-452-5

Illegal, Unreported, and Unregulated (IUU) Fishing: Impacts, Costs, and Countermeasures
Lorenzo Scherpiani and Stefano Rea
ISBN: 979-8-89530-084-8
eBook ISBN: 979-8-89530-191-3

Crayfish: Evolution, Habitat and Conservation Strategies
Felipe Bezerra Ribeiro (Editor)
ISBN: 978-1-53616-941-6
eBook ISBN: 978-1-53616-942-3

Sexual Plasticity and Gametogenesis in Fishes
Balasubramanian Senthilkumaran (Editor)
ISBN: 978-1-62618-848-8
eBook ISBN: 978-1-62618-879-2

More information about this series can be found at
https://novapublishers.com/product-category/series/fish-fishing-and-fisheries/

Gordon B. Maddox

Ongoing Efforts to Combat Illegal, Unreported, and Unregulated (IUU) Fishing

Copyright © 2026 by Nova Science Publishers, Inc.

All rights reserved. No part of this book may be reproduced, stored in a retrieval system or transmitted in any form or by any means: electronic, electrostatic, magnetic, tape, mechanical photocopying, recording or otherwise without the written permission of the Publisher.

We have partnered with Copyright Clearance Center to make it easy for you to obtain permissions to reuse content from this publication. Simply navigate to this publication's page on Nova's website and locate the "Get Permission" button below the title description. This button is linked directly to the title's permission page on copyright.com. Alternatively, you can visit copyright.com and search by title, ISBN, or ISSN.

For further questions about using the service on copyright.com, please contact:
Copyright Clearance Center
Phone: +1-(978) 750-8400   Fax: +1-(978) 750-4470   E-mail: info@copyright.com.

NOTICE TO THE READER

The Publisher has taken reasonable care in the preparation of this book, but makes no expressed or implied warranty of any kind and assumes no responsibility for any errors or omissions. No liability is assumed for incidental or consequential damages in connection with or arising out of information contained in this book. The Publisher shall not be liable for any special, consequential, or exemplary damages resulting, in whole or in part, from the readers' use of, or reliance upon, this material. Any parts of this book based on government reports are so indicated and copyright is claimed for those parts to the extent applicable to compilations of such works.

Independent verification should be sought for any data, advice or recommendations contained in this book. In addition, no responsibility is assumed by the Publisher for any injury and/or damage to persons or property arising from any methods, products, instructions, ideas or otherwise contained in this publication.

The Publisher assumes no responsibility for any statements of fact or opinion expressed in the published contents.

This publication is designed to provide accurate and authoritative information with regard to the subject matter covered herein. It is sold with the clear understanding that the Publisher is not engaged in rendering legal or any other professional services. If legal or any other expert assistance is required, the services of a competent person should be sought. FROM A DECLARATION OF PARTICIPANTS JOINTLY ADOPTED BY A COMMITTEE OF THE AMERICAN BAR ASSOCIATION AND A COMMITTEE OF PUBLISHERS.

Additional color graphics may be available in the e-book version of this book.

Library of Congress Cataloging-in-Publication Data

ISBN: 979-8-89530-858-5 (Softcover)
ISBN: 979-8-89530-959-9 (eBook)

Published by Nova Science Publishers, Inc. † New York

Contents

Preface		vii
Chapter 1	**Understanding Illegal, Unreported, and Unregulated Fishing**	1
	National Oceanic and Atmospheric Administration	
Chapter 2	**Illegal, Unreported, and Unregulated (IUU) Fishing: Frequently Asked Questions**	5
	Caitlin Keating-Bitonti and Anthony R. Marshak	
Chapter 3	**Combating Illegal Fishing: Clear Authority Could Enhance U.S. Efforts to Partner with Other Nations at Sea**	57
	United States Government Accountability Office	
Chapter 4	**Combating Illegal Fishing: Better Information Sharing Could Enhance U.S. Efforts to Target Seafood Imports for Investigation**	97
	United States Government Accountability Office	
Chapter 5	**National 5-Year Strategy for Combating Illegal, Unreported, and Unregulated Fishing**	125
	Janet Coit and Richard W. Spinrad	
Chapter 6	**Tackling Illicit Fishing at Sea and Ports Before It Ends Up on Your Plate**	159
	United States Government Accountability Office	
Chapter 7	**Guardians of the Sea: Examining Coast Guard Efforts in Drug Enforcement, Illegal Migration, and IUU Fishing**	163
	Caitlin Keating-Bitonti	

Chapter 8	**Maritime SAFE Act Interagency Working Group on IUU Fishing Priority Regions**......................177	
	Maritime SAFE Act Interagency Working Group	
Index	181	

Preface

Illegal, unreported and unregulated (IUU) fishing violates both national and international fishing regulations. IUU fishing is a global problem that threatens marine ecosystems and sustainable fisheries. It also threatens our economic security and the natural resources critical to global food security, and it puts law-abiding fishermen and seafood producers in the United States and abroad at a disadvantage. This book addresses issues related to IUU fishing.

Chapter 1

Understanding Illegal, Unreported, and Unregulated Fishing[*]

National Oceanic and Atmospheric Administration

IUU fishing is a global problem threatening ocean ecosystems and sustainable fisheries. Learn more about IUU fishing and NOAA's role in combating these activities.

What Is Illegal, Unreported, and Unregulated Fishing?

Illegal, unreported, and unregulated fishing activities violate both national and international fishing regulations. IUU fishing is a global problem that threatens ocean ecosystems and sustainable fisheries. It also threatens our economic security and the natural resources that are critical to global food security, and it puts law-abiding fishermen and seafood producers in the United States and abroad at a disadvantage.

Illegal fishing refers to fishing activities conducted in contravention of applicable laws and regulations, including those laws and rules adopted at the regional and international level.

Unreported fishing refers to fishing activities that are not reported or are misreported to relevant authorities in contravention of national laws and regulations or reporting procedures of a relevant regional fisheries management organization.

---

[*] This is an edited, reformatted and augmented version of National Oceanic and Atmospheric Administration Publication.

In: Ongoing Efforts to Combat Illegal, Unreported …
Editor: Gordon B. Maddox
ISBN: 979-8-89530-858-5
© 2026 Nova Science Publishers, Inc.

Unregulated fishing occurs in areas or for fish stocks for which there are no applicable conservation or management measures and where such fishing activities are conducted in a manner inconsistent with State responsibilities for the conservation of living marine resources under international law. Fishing activities are also unregulated when occurring in an RFMO-managed area and conducted by vessels without nationality, or by those flying a flag of a State or fishing entity that is not party to the RFMO in a manner that is inconsistent with the conservation measures of that RFMO. Learn more about RFMOs.

How Does the United States Work with Other Countries to Combat IUU Fishing?

The United States, as both a major consumer and major producer of seafood, is leading efforts to ensure that its high demand for imported seafood does not create incentives for IUU fishing activities. Combating IUU fishing is a top priority for our nation, and NOAA Fisheries is leading these efforts by:

- Working with other fishing nations through regional fisheries bodies and international partnerships. Together, we are strengthening enforcement and data collection programs aimed at detecting, deterring, and eliminating IUU fishing.
- Implementing measures that restrict port entry and access to port services for vessels included on the IUU lists of international fisheries organizations with U.S. membership. Learn about the Port State Measures Agreement.
- Identifying countries that have fishing vessels engaged in IUU fishing activities under the Magnuson-Stevens Reauthorization Act. Learn about the Biennial Reports to Congress on IUU Fishing, Bycatch and Shark Catch.
- Implementing reporting and recordkeeping measures to prevent IUU-caught seafood from entering the United States. Learn more about the Seafood Monitoring Program.
- Supporting capacity-building and technical assistance workshops that provide the tools, resources, information, and skills to solve IUU issues, combat IUU fishing, and promote sustainable seafood practices. Learn more about capacity building.

What Are Some Examples of IUU Fishing Activities?

IUU fishing includes:

- Fishing without a license or quota for certain species.
- Failing to report catches or making false reports.
- Keeping undersized fish or fish that are otherwise protected by regulations.
- Fishing in closed areas or during closed seasons, and using prohibited fishing gear.
- Conducting unauthorized transshipments (e.g., transfers of fish) to cargo vessels.

Who Is Most Affected by IUU Fishing?

IUU fishing poses a direct threat to food security and socioeconomic stability in many parts of the world. Developing countries that depend on fisheries for food security and export income are most at risk from IUU fishing. For example, total catches in West Africa are estimated to be 40 percent higher than reported catches. Many crew members on IUU fishing vessels are from poor and underdeveloped parts of the world, and they often work in unsafe conditions.

What Are the Economic Losses Caused by IUU Fishing?

The inherent nature of illegal, unreported, and unregulated fishing makes it difficult to accurately quantify the full global economic impacts resulting from these activities. But there is little disagreement that it is in the billions, or even tens of billions, of dollars each year. Various studies over the years have assessed regional levels of IUU fishing and estimated global losses, but such estimates are based on data that are now many years old. The United Nation's Food and Agricultural Organization is currently developing regional IUU estimate methodologies that can be regularly updated. Implementing the UN's action plan recommendations will help gauge the actual level of activity and impacts so that they may be appropriately addressed.

How Does IUU Fishing Affect the Seafood Industry and U.S. Consumers?

Fishermen and companies that engage in IUU fishing circumvent conservation and management measures, avoid the operational costs associated with sustainable fishing practices, and may derive economic benefit from exceeding harvesting limits. As a result, their illegally caught products can provide unfair competition in the marketplace for law-abiding fishermen and seafood industries.

Additionally, because the United States imports most of its seafood, IUU fishing creates an imbalance of lower quality products in the seafood market while simultaneously hindering the sustainability of global marine resources.

How Can I Learn More About What the United States Is Doing to Combat IUU Fishing?

Combating IUU fishing is a top priority for the United States, and communication, collaboration and strategic coordination will be key in bringing about tangible results. The U.S. Interagency Working Group on IUU Fishing brings together twenty-one agencies for an integrated, federal government-wide response to IUU fishing globally. The efforts of the U.S. Interagency Working Group on IUU Fishing now sit at the heart of our government's coordination on tackling IUU fishing practices and setting the conditions where IUU fishing is neither accepted nor commonplace in the future.

Chapter 2

Illegal, Unreported, and Unregulated (IUU) Fishing: Frequently Asked Questions*

**Caitlin Keating-Bitonti
and Anthony R. Marshak**

Summary

Over the past two decades, successive U.S. Administrations and Congresses have recognized that illegal, unreported, and unregulated (IUU) fishing threatens national, regional, and global security and have acted to combat such fishing activities. IUU fishing generally refers to fishing activities—occurring both in coastal nation jurisdictional waters and in international waters (i.e., the high seas)—that violate national laws or international fisheries conservation and management measures. Some IUU fishing vessels also may engage in other transnational crimes, such as human trafficking and/or labor abuses, as well as the smuggling of drugs, arms, and wildlife. IUU fishing may have several co-occurring consequences that range from harming legitimate (i.e., law-abiding) commercial fishing to undermining scientifically informed fisheries management. IUU fishing also can threaten local and regional food and economic security.

Congress has passed several laws directly or indirectly addressing IUU fishing within U.S. waters and/or on the high seas. Some of these laws focus on addressing the impacts of marine biodiversity loss associated with IUU fishing (e.g., High Seas Driftnet Fishing Moratorium Protection Act; Title VI of P.L. 104-43, and the Magnuson-Stevens Fishery Conservation and Management Reauthorization Act of 2006;

---

* This is an edited, reformatted and augmented version of Congressional Research Service Publication No. R48215, dated October 8, 2024.

In: Ongoing Efforts to Combat Illegal, Unreported …
Editor: Gordon B. Maddox
ISBN: 979-8-89530-858-5
© 2026 Nova Science Publishers, Inc.

Title IV, §§401-403, of P.L. 109-479). Other U.S. laws address the law enforcement aspects of IUU fishing, such as the Maritime Security and Fisheries Enforcement Act (commonly known as the Maritime SAFE Act; Division C, Title XXXV, Subtitle C, of P.L. 116-92). The Maritime SAFE Act provided a whole-of- government approach to address IUU fishing globally and established the Interagency Working Group on IUU Fishing. In addition, Congress directed the Secretary of Commerce, through the Consolidated Appropriations Act, 2018 (P.L. 115-141), to implement the U.S. Seafood Import Monitoring Program (SIMP) to prevent imported IUU fish and fish products from entering U.S. commerce.

The Food and Agriculture Organization of the United Nations (FAO) provides an international framework to address IUU fishing globally and implements several international fisheries legal instruments. The United States is a party to numerous FAO and other international agreements aimed at curbing or preventing IUU fishing, including the 1995 UN Fish Stocks Agreement, the Port State Measures Agreement, the Agreement to Prevent Unregulated High Seas Fisheries in the Central Arctic Ocean, and the World Trade Organization Agreement on Fisheries Subsidies. The United States also is a member of several *regional fisheries management organizations* (RFMOs), which are international fishery management bodies established to conserve and manage transboundary fish stocks (i.e., fish that move across maritime zones) and fisheries on the high seas. In 2019, under the Maritime SAFE Act, Congress directed the Secretary of State, in consultation with the Secretary of Commerce, to coordinate with RFMOs to enhance regional responses to IUU fishing. Because most IUU activities occur outside of U.S. jurisdiction, the U.S. government has taken several actions—through international agreements, organizations, and trade—to influence the behavior of foreign fishing fleets. Several federal agencies, including the Department of Defense, the Department of State, the National Oceanic and Atmospheric Administration, and the U.S. Coast Guard, participate in various efforts to combat IUU fishing on the high seas and in the exclusive economic zones (i.e., the waters 200 nautical miles seaward from the shoreline under the jurisdiction of coastal nations) of partner nations. Such efforts include establishing strategic partnerships, improving enforcement tools, identifying and sharing information about vessels participating in IUU fishing, and assisting partner nations in developing and maintaining their own capacities to counter IUU fishing, among others.

To address IUU fishing, Congress may consider several policy options. For example, Congress may consider including other illicit activities that often occur in the seafood sector (e.g., human trafficking, forced labor) in the U.S. definition of IUU fishing. As another example, Congress may consider directing certain federal agencies to enhance transparency and traceability across the U.S. seafood supply chain, as well as to expand SIMP to include all species imported by the United States, to help keep seafood derived from IUU fishing out of the U.S. marketplace. Congress also may consider whether more resources and greater diplomatic support could help in the coordination of fishery management, including in regions currently without RFMOs. In addition, Congress may examine whether sufficient support and resources have been dedicated to enforcement efforts to counter IUU fishing activities that may include capacity-building assistance to coastal nations and joint efforts, such as shiprider agreements, among other potential considerations.

Fisheries resources are a major component of the U.S. economy. In 2022, the United States imported 17% of the world import value of aquatic animal products ($32 billion), was the largest individual importing country, and was the fifth-largest wild seafood producer, according to a 2024 report by the Food and Agriculture Organization of the United Nations (FAO).[1] The United States has shown a strong interest in the conservation and sustainable management of fisheries resources, both within areas of national jurisdiction and on the high seas (international waters, which cover approximately 45% of the planet).[2] U.S. and international efforts to manage fisheries are weakened by illegal, unreported, and unregulated (IUU) fishing.[3]

IUU fishing is an ongoing, multifaceted global issue that occurs both within areas of national jurisdiction and on the high seas.[4] Under the United Nations Convention on the Law of the Sea (UNCLOS), coastal nations determine allowable catches and promote optimal resource use within their exclusive economic zones (EEZs), waters 200 nautical miles seaward from their shorelines.[5] However, many living resources move between waters under

---

[1] Food and Agriculture Organization of the United Nations (FAO), The State of World Fisheries and Aquaculture: Blue Transformation in Action, 2024, pp. 30, 90. Hereinafter FAO, State of World Fisheries and Aquaculture, 2024.

[2] U.S. Department of State, "Illegal, Unreported, and Unregulated Fishing," https://www.state.gov/key-topics-office-of-marine-conservation/illegal-unreported-and-unregulated-fishing/. Hereinafter U.S. Department of State, "Illegal, Unreported, and Unregulated Fishing."

[3] FAO, "Illegal, Unreported, Unregulated (IUU) Fishing," https://www.fao.org/iuu-fishing/en/.

[4] Given the multifaceted nature of illegal, unreported, and unregulated (IUU) fishing, several CRS experts cover aspects of the topic. For information about the experts, see CRS Report R47859, Illegal, Unreported, and Unregulated (IUU) Fishing: CRS Experts, coordinated by Caitlin Keating-Bitonti.

[5] Article 57 of the United Nations Convention on the Law of the Sea (UNCLOS) defines the breadth of the exclusive economic zone (EEZ). According to UNCLOS, within its EEZ, a coastal nation has the right to explore, exploit, conserve, and manage living and nonliving

national jurisdiction—internal waters (e.g., the Chesapeake Bay), the territorial sea (12 nautical miles seaward from a coastal nation's shoreline), the EEZ—and the high seas. For example, straddling fish stocks can be found both within a nation's EEZ and in the adjacent high seas, and highly migratory fish stocks regularly travel long distances through high seas areas and areas under national jurisdiction (Figure 1). Activities associated with IUU fishing affect the ocean ecosystem and the sustainable management of living marine resources.[6]

The vast majority (95%) of global marine fish catch occurs within EEZs.[7] Coastal nations with limited maritime patrol and enforcement capabilities are particularly susceptible to IUU fishing occurring within their EEZs. National fishers, as well as fishers from neighboring coastal nations, may engage in IUU fishing within EEZs. Other fishers may travel great distances across the ocean, crossing jurisdictional boundaries to engage in IUU fishing in another coastal nation's EEZ. Some fishers participate in IUU fishing activities on the high seas because high seas areas often have limited enforcement and patchy regulation (i.e., some areas of the high seas do not have fisheries management guidelines, regulations, or associated enforcement).

This report addresses 14 frequently asked questions related to IUU fishing.

---

natural resources of the seabed and subsoil and the above-water column. The United States has not ratified UNCLOS but generally abides by certain provisions of the convention's terms, as dictated by Presidential Proclamation 5030. See "Proclamation 5030: Exclusive Economic Zone of the United States of America," 48 Federal Register 10605 (March 10, 1983).

[6] National Oceanic and Atmospheric Administration (NOAA), National Marine Fisheries Service (NMFS), Report to Congress: Improving International Fisheries Management, August 2023, p. 10; and NOAA, NMFS, Report to Congress. Report on the Seafood Import Monitoring Program—FY2023, 2024, pp. 1-19 (hereinafter NOAA, NMFS, Report on the Seafood Import Monitoring Program—FY2023).

[7] Liam Campling et al., "A Geopolitical-Economy of Distant Water Fisheries Access Arrangements" npj Ocean Sustainability, vol. 3, no. 26 (2024), p. 1. Hereinafter Campling et al., "Geopolitical-Economy of Distant Water Fisheries."

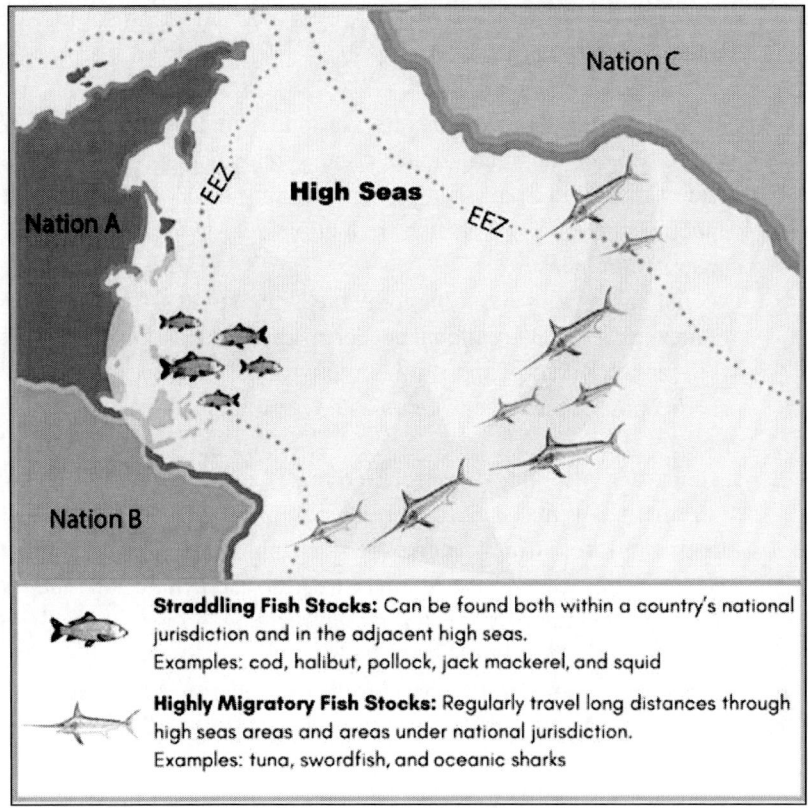

Source: Congressional Research Service, modified from United Nations, Division for Ocean Affairs and the Law of the Sea, "UNFSA Overview: The United Nations Fish Stocks Agreement," https://www.un.org/oceancapacity/ UNFSA.

Notes: EEZ = exclusive economic zone. The EEZ separates waters under national jurisdiction (the first 200 nautical miles seaward from a coastal nation's shoreline) and the high seas (international waters).

Figure 1. Straddling and Highly Migratory Fish Stocks.

What Is IUU Fishing?

IUU fishing generally refers to fishing activities that violate national laws or international fisheries conservation and management measures. Congress directed the Secretary of Commerce to publish a definition of the term illegal, unreported, or unregulated fishing as part of IUU- related provisions in the Magnuson-Stevens Fishery Conservation and Management Reauthorization

Act of 2006 (Title IV, §§401-403, of P.L. 109-479).[8] The act also provided guidance that the definition include

   a. fishing activities that violate conservation and management measures required under an international fishery management agreement to which the United States is a party, including catch limits or quotas, capacity restrictions, bycatch reduction requirements, and shark conservation measures;
   b. overfishing of fish stocks shared by the United States, for which there are no applicable international conservation or management measures or in areas with no applicable international fishery management organization or agreement, that has adverse impacts on such stocks; and
   c. fishing activity that has an adverse impact on seamounts, hydrothermal vents, and cold water corals located beyond national jurisdiction, for which there are no applicable conservation or management measures or in areas with no applicable international fishery management organization or agreement.[9]

In 2011, the National Oceanic and Atmospheric Administration's (NOAA's) National Marine Fisheries Service (NMFS) defined the term illegal, unreported, or unregulated fishing in a final rule.[10] Since then, NMFS has made changes to the definition.[11] The current definition appears in 50 C.F.R. §300.201 as follows.

Illegal, unreported, or unregulated (IUU) fishing means:

1) In the case of parties to an international fishery management agreement to which the United States is a party, fishing activities that violate conservation and management measures required under an international fishery management agreement to which the United

---

[8] Codified at 16 U.S.C. §1826j(e)(2). The act amended provisions included in the High Seas Driftnet Fishing Moratorium Protection Act (Title VI of P.L. 104-43), among other changes.
[9] 16 U.S.C. §1826j(e)(3).
[10] NOAA, NMFS, "High Seas Driftnet Fishing Moratorium Protection Act; Identification and Certification Procedures to Address Illegal, Unreported, and Unregulated Fishing Activities and Bycatch of Protected Living Marine Resources," 76 *Federal Register* 2024, January 12, 2011.
[11] NOAA, NMFS, "High Seas Driftnet Fishing Moratorium Protection Act; Identification and Certification Procedures to Address Shark Conservation," 78 *Federal Register* 3338-3346, January 16, 2013.

States is a party, including but not limited to catch limits or quotas, capacity restrictions, bycatch reduction requirements, shark conservation measures, and data reporting;
2) In the case of non-parties to an international fishery management agreement to which the United States is a party, fishing activities that would undermine the conservation of the resources managed under that agreement;
3) Overfishing of fish stocks shared by the United States, for which there are no applicable international conservation or management measures, or in areas with no applicable international fishery management organization or agreement, that has adverse impacts on such stocks; or,
4) Fishing activity that has a significant adverse impact on seamounts, hydrothermal vents, cold water corals and other vulnerable marine ecosystems located beyond any national jurisdiction, for which there are no applicable conservation or management measures or in areas with no applicable international fishery management organization or agreement.
5) Fishing activities by foreign flagged vessels in U.S. waters without authorization of the United States.

The 2001 FAO International Plan of Action to Prevent, Deter, and Eliminate Illegal, Unreported, and Unregulated Fishing (IPOA-IUU) provides a definition for IUU fishing,[12] as well as a "toolbox" of voluntary measures that countries and fishers can take to address IUU fishing.[13] The Agreement on Port State Measures to Prevent, Deter, and Eliminate Illegal, Unreported, and Unregulated Fishing uses the same IUU fishing definition as the IPOA-IUU.[14]

---

[12] The International Plan of Action to Prevent, Deter, and Eliminate Illegal, Unreported, and Unregulated Fishing (IPOA-IUU) is a voluntary instrument under FAO's Code of Conduct for Responsible Fisheries. The document contains separate definitions for *illegal fishing*, *unreported fishing*, and *unregulated fishing*. FAO, *International Plan of Action to Prevent, Deter, and Eliminate Illegal, Unreported, and Unregulated Fishing,* Rome, Italy, 2001, pp. 2-3.

[13] The National Plan of Action of the United States of America to Prevent, Deter, and Eliminate Illegal, Unregulated, and Unreported Fishing "is organized along the same lines as the [IPOA-IUU]." U.S. Department of State, *National Plan of Action of the United States of America to Prevent, Deter, and Eliminate Illegal, Unreported, and Unregulated Fishing*, p. 2, https://2001-2009.state.gov/documents/organization/43101.pdf.

[14] FAO, Agreement on Port State Measures to Prevent, Deter, and Eliminate Illegal, Unreported, and Unregulated Fishing, Rome, Italy, June 20, 2012. Hereinafter FAO, Port State Measures Agreement. For more information, see the "Port State Measures Agreement" section of this report.

The U.S. definition for IUU fishing is consistent with the terminology used in the IPOA-IUU definition.

Neither the United States' definition nor the international definition for IUU fishing includes human trafficking or other illicit activities that are commonly associated with IUU fishing. Some stakeholders contend the definition for IUU fishing should be broadened to include illicit fisheries-related crimes and crimes committed in the context of the fisheries sector, such as human trafficking and forced labor.[15] In a March 2024 letter to President Biden, some Members of Congress noted that the NOAA definition does not encompass human and labor rights abuses and stated that updating the definition would align with international standards.[16]

What Drives IUU Fishing?

Several factors drive fishers to participate in IUU fishing activities, such as the profitability of the seafood trade, limited enforcement and patchy regulation of the high seas, government fisheries subsidies, and the ability to use flags of convenience. Distant water fishing (DWF) fleets are operated by firms fishing in areas outside the jurisdiction where ownership (or registration) is held and beyond the area(s) adjacent to the natural shoreline of that jurisdiction.[17] The top five DWF fleets—China (38%), Taiwan (22%), Japan (10%), South Korea (10%), and Spain (10%)— account for 90% of all DWF and primarily operate in the Indo-Pacific region and off the coasts of East and West Africa and South America.[18]

Profitability. Wild-caught and aquaculture seafood products represent some of the most internationally traded food commodities. In 2022,

---

[15] See, for example, Julio Jorge Urbina, "Towards an International Legal Definition of the Notion of Fisheries Crime," *Marine Policy*, vol. 144 (October 2022), pp. 1-6, see p. 2; Vasco Becker-Weinberg, "Time to Get Serious About Combating Forced Labour and Human Trafficking in Fisheries," *International Journal of Marine and Coastal Law*, vol. 36 (2021), pp. 88-113; and Mary Mackay et al., "The Intersection Between Illegal Fishing, Crimes at Sea, and Social Well-Being," *Frontiers in Marine Science*, vol. 7 (2020), 589000, pp. 1-9.

[16] Letter from Congress of the United States to the Honorable Joseph R. Biden Jr., President of the United States, March 11, 2024, p. 3, https://democrats-naturalresources.house.gov/imo/media/doc/2024-03-11_moc_letter_to_president_biden_re_iuu_fishing.pdf.

[17] Campling et al., "Geopolitical-Economy of Distant Water Fisheries," pp. 2-3.

[18] Stimson Center, *Shining a Light: The Need for Transparency Across Distant Water Fishing*, 2019, p. 2, https://www.stimson.org/wp-content/files/file-attachments/Stimson%20Distant%20Water%20Fishing%20Report.pdf. Hereinafter Stimson Center, *Shining a Light*.

approximately 38% of wild-caught and aquaculture products entered into international trade, generating hundreds of billions of dollars in revenue.[19] To take advantage of the profitable global fisheries market, fishers may engage in IUU fishing to avoid the operational costs associated with sustainable fisheries management.[20] Experts contend that IUU fishing harms legitimate (i.e., law-abiding) commercial fishers and disrupts efforts toward sustainable fishing practices, while resulting in global economic losses of an estimated tens of billions of dollars each year for legitimate seafood producers.[21]

Limited Enforcement and Patchy Regulations on the High Seas. The high seas cover approximately 45% of the planet. The large area, coupled with limited enforcement and patchy regulation, enables some fishers to engage in IUU fishing activities on the high seas.[22] Under the Agreement for the Implementation of the Provisions of the United Nations Convention on the Law of the Sea of 10 December 1982 Relating to the Conservation and Management of Straddling Fish Stocks and Highly Migratory Fish Stocks (commonly known as the 1995 UN Fish Stocks Agreement), countries party to the agreement are obligated to regulate "the activities of vessels flying their flag which fish for [straddling fish stocks and highly migratory fish stocks] on the high seas."[23]

Fisheries Subsidies.[24] The World Trade Organization (WTO) defines *subsidies* as a financial contribution made by a government or any public body that confers a benefit. Economists generally consider subsidies to be trade distorting. Since the 1990s, multilateral organizations, such as FAO, and other stakeholders have raised concerns about how fisheries subsidies contribute to overcapacity and overfishing, and cause economic and environmental

---

[19] FAO, *State of World Fisheries and Aquaculture*, 2024, pp. 9, 82-83.
[20] U.S. Department of State, "Illegal, Unreported, and Unregulated Fishing."
[21] For example, see U.S. Department of State, "Illegal, Unreported, and Unregulated Fishing"; and National Intelligence Council (NIC), *Global Implications of Illegal, Unreported, and Unregulated (IUU) Fishing*, NIC WP 2016-02, September 19, 2016, p. 5 (hereinafter, NIC, *Global Implications of IUU Fishing*).
[22] Stimson Center, *Shining a Light*, p. 2. For more information, see the "How Can Regional Fisheries Management Organizations Deter IUU Fishing?" section of this report.
[23] Article 7 of the Agreement for the Implementation of the Provisions of the United Nations Convention on the Law of the Sea of 10 December 1982 Relating to the Conservation and Management of Straddling Fish Stocks and Highly Migratory Fish Stocks, https://www.un.org/depts/los/convention_agreements/texts/fish_stocks_agreement/CONF164_37.htm. Hereinafter 1995 UN Fish Stocks Agreement. For more information, see the "1995 UN Fish Stocks Agreement" section of this report.
[24] Liana Wong, Analyst in International Trade and Finance, contributed to this section on fisheries subsidies.

impacts.[25] Subsidies provide capital to fisheries to expand fishing fleets and increase capacity to fish.

Decreasing fisheries subsidies could make some IUU fishing operations unprofitable, potentially decreasing fishing effort and improving conservation and management efforts.[26]

Flag of Convenience. Vessels must be registered with a single country even if they operate on the high seas. A *flag of convenience* vessel is one that flies the flag of a country other than the country of vessel ownership.[27] Registering for a flag of convenience can be attractive to some vessel owners, especially if the country of the flag of convenience has low registration fees, low or no taxes, relaxed labor laws, and limited high seas enforcement capabilities in comparison to the nation of ownership.[28] Many operators involved in IUU fishing activities register their vessel or fleet in a nation that lacks the capacity and resources for effective monitoring, control, and surveillance.[29] Vessel operators also may seek to register in a country with "limited interest or ability to enforce fishing or labor-related laws."[30] Flag-of-convenience vessels can be challenging to track because some frequently change their name, ownership, and country of registration.

---

[25] FAO, *State of World Fisheries and Aquaculture*, 2024, p. xxv; Ussif Rashid Sumaila et al., "Updated Estimates and Analysis of Global Fisheries Subsidies," *Marine Policy*, vol. 109 (November 2019), pp. 1-11; and World Bank, *The Sunken Billions Revisited: Progress and Challenges in Global Marine Fisheries,* 2017, pp. 1-4.

[26] For more information see the "World Trade Organization Agreement on Fisheries Subsidies" section of this report. NOAA defines *fishing effort* as "the amount of fishing gear of a specific type used on the fishing grounds over a given unit of time (e.g., hours trawled per day, number of hooks set per day, number of hauls of a beach seine per day). When two or more kinds of gear are used, the respective efforts must be adjusted to some standard type before being added." U.S. Department of Commerce, NOAA, *NOAA Fisheries Glossary*, NOAA Technical Memorandum NMFS/F-SPO-69, June 2006, p. 17, https://repository.library.noaa.gov/view/noaa/12856.

[27] For example, see International Transport Workers' Federation, "Flags of Convenience," https://www.itfglobal.org/ en/sector/seafarers/flags-convenience.

[28] Ibid.

[29] FAO, "Illegal, Unreported, Unregulated (IUU) Fishing," https://www.fao.org/iuu-fishing/en/.

[30] U.S. Department of Commerce and U.S. Department of State, *Human Trafficking in the Seafood Supply Chain: Section 3563 of the National Defense Authorization Act of Fiscal Year 2020 (P.L. 116-92)*, Report to Congress, 2020, https://media.fisheries.noaa.gov/2020-12/DOSNOAAReport_HumanTrafficking.pdf. Hereinafter Departments of Commerce and State, *Human Trafficking in the Seafood Supply Chain*.

What Are Examples of IUU Fishing Activities?

IUU fishing generally refers to fishing activities that violate national laws or international fisheries conservation and management measures. For example, licensed vessels may operate in violation of national laws within coastal nations' EEZs, misreport their harvests, or not comply with *regional fisheries management organization* (RFMO) measures in high seas areas, among other violations (*Figure 2*).[31] As another example, unlicensed vessels may operate illegally within EEZs or participate in unregulated fishing (i.e., fishing activities in areas where there are no applicable conservation or management areas, such as RFMOs).[32]

What Are the Consequences of IUU Fishing?

IUU fishing may have several co-occurring consequences that range from harming legitimate commercial fishers to exacerbating overfishing, inhibiting fisheries research, and undermining fisheries management. IUU fishing also can threaten food security. The threats that IUU fishing activities pose to local and national economies may increase tensions within and between countries and may contribute to broader geopolitical conflicts. In addition, IUU fishing may impact the seafood industry.[33]

Harm Legitimate Commercial Fishers. IUU fishing adversely impacts legitimate commercial fishers.[34] Vessels conducting IUU fishing avoid operational costs by not complying with regulatory requirements such as gear restrictions, closed areas, or harvest limits. The decline of common or shared stocks because of illegal fishers may lead to lower harvests for legitimate fishers. Those fishing legally may be harmed by lower catch rates and higher associated fishing costs.

Exacerbate Overfishing and Undermine Fisheries Management. IUU fishing activities can contribute to overfishing, deplete protected living marine

---

[31] A *regional fisheries management organization* (RFMO) is an international fishery management body established to conserve and manage fish stocks that move across maritime zones. For more information, see the "How Can Regional Fisheries Management Organizations Deter IUU Fishing?" section of this report.
[32] See NIC, *Global Implications of IUU Fishing*, p. 5.
[33] For more information, see the "How Does IUU Fishing Impact the Seafood Industry?" section of this report.
[34] For example, see U.S. Department of State, "Illegal, Unreported, and Unregulated Fishing"; and NIC, *Global Implications of IUU Fishing*, p. 5.

resources, and diminish efforts to assess and manage marine populations.[35] Fishers participating in IUU fishing activities may underreport or not report their catch, thereby inhibiting efforts by scientists and fisheries managers to assess fish population dynamics and set catch limits.

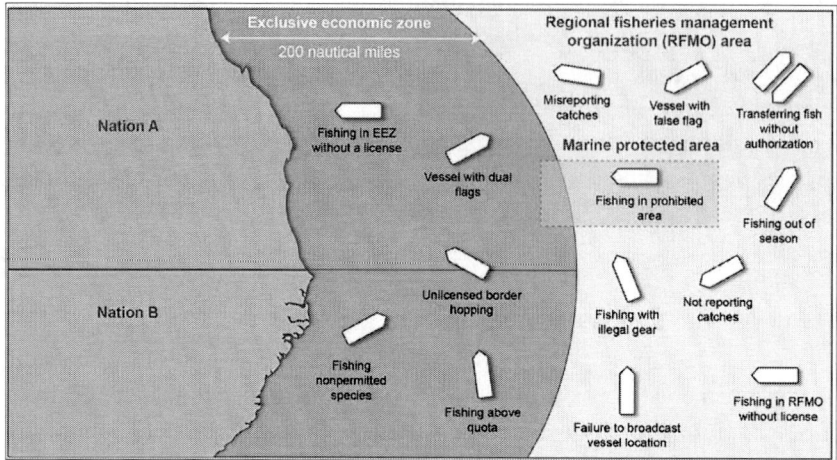

Source: U.S. Government Accountability Office, Combatting Illegal Fishing: Better Information Sharing Could Enhance U.S. Efforts to Target Seafood Imports for Investigation, GAO-23-105643, May 2023, see, p. 6.

Notes: EEZ = exclusive economic zone. The green and brown areas represent two coastal nations, and the dark blue area represents the two nations' adjacent EEZs (i.e., 200-nautical-mile area under their respective national jurisdictions). The light blue area represents an area of the high seas (international waters) under the management of a regional fisheries management organization (RFMO). An RFMO is an international fishery management body established to conserve and manage transboundary fish stocks (i.e., fish that move across maritime zones). RFMOs also manage fishing activities occurring within a specific geographic region of the high seas. The area delineated by the hashed rectangle represents a marine protected area that spans a portion of nation A's EEZ and an adjacent area of the high seas. Some marine protected areas prohibit certain activities, such as commercial fishing.

Figure 2. Examples of Illegal, Unreported, and Unregulated Fishing Activities.

Overfishing of a fish stock in one area can affect the stock condition of that species in other parts of its range.[36] IUU fishing presents wide-ranging

---

[35] For example, see FAO, State of World Fisheries and Aquaculture, 2024, p. 146.
[36] For example, see Ines Haberle et al., "Fish Condition as an Indicator of Stock Status: Insights from Condition Index in a Food-Limiting Environment," *Fish and Fisheries*, vol. 24, no. 4

management and conservation challenges at both national and international levels. International cooperation is necessary to manage many fish stocks, because they move among different national zones of jurisdiction and the high seas (*Figure 1*).

Threaten Food Security.[37] IUU fishing's combined effects of overfishing and disruption of traditional seafood markets may contribute to food scarcity. IUU fishing presents a threat to food security and socioeconomic stability in many countries, especially in developing nations that depend on fisheries for food and export income.[38] According to FAO, aquaculture and fisheries combined accounted for approximately 17% of total animal-source protein consumed globally in 2017, with these values typically greater in developing nations of coastal Africa, South Asia, Oceania, and parts of the Caribbean.[39] For some countries in western Africa and South Asia, fish contributes 50% or more of their residents' total animal protein intake.[40] In several of these regions, foreign vessels, including DWF vessels, fish illegally and contribute to overexploitation.[41] As demand for marine protein sources is anticipated to increase,[42] experts note that food derived from the global ocean

---

(2023), pp. 567-581; and Helen F. Yan et al., "Overfishing and Habitat Loss Drive Range Contraction of Iconic Marine Fishes to Near Extinction," *Science Advances*, vol. 7, no. 7 (2021), eabb6026, pp. 1-10.

[37] FAO defines *food security* at the individual, household, national, regional, and global levels as existing "when all people, at all times, have physical and economic access to sufficient, safe and nutritious food to meet their dietary needs and food preferences for an active and healthy life." FAO, *Rome Declaration on World Food Security and World Food Summit Plan of Action*, paper presented at the Word Food Summit, Rome, Italy, 1996.

[38] NOAA, NMFS, "Understanding Illegal, Unreported, and Unregulated Fishing," https://www.fisheries.noaa.gov/insight/understanding-illegal-unreported-and-unregulated-fishing. Hereinafter NOAA, NMFS, "Understanding Illegal, Unreported, and Unregulated Fishing."

[39] FAO, *The State of World Fisheries and Aquaculture 2020: Sustainability in Action*, 2020, p. 67.

[40] Ibid.

[41] See, for example, Yimin Ye et al., "Increasing the Contribution of Africa's Fisheries to Food Security Through Improved Management," *Food Security*, vol. 16, no. 2 (2024), pp. 455-470; Nudrin Kasim and Aris Widagdo, "Combatting Illegal, Unreported, and Unregulated (IUU) Fishing in Indonesia," *Aquaculture, Aquarium, Conservation and Legislation*, vol. 12, no. 6 (2019), pp. 2243-2251; and Matti Kohonen and Alfonso Daniels, "Ocean Economy at Risk: Rise of Distant Water Fleets and Financial Secrecy," *Development*, vol. 66 (2023), pp. 76-83.

[42] Det Norske Veritas (DNV), *Seafood Forecast: Ocean's Future to 2050*, Høvik, Norway, 2024, pp. 1-69, https://www.dnv.com/Publications/seafood-forecast-250243/.

is increasingly important for food security and that IUU fishing undermines sustainable food production.[43]

How Does IUU Fishing Impact the Seafood Industry?

IUU fishing causes direct economic impacts to the global seafood industry (including the industries of particular nations) and to U.S. domestic seafood in general. Some Members of Congress have asked specific agencies to provide information about how the agency is responding to the threat that IUU fishing poses to domestic seafood production,[44] in reference to a U.S. International Trade Commission report that found nearly 11% of all U.S. seafood imports, equal to $2.4 billion annually, are derived from IUU fishing.[45] Multiple studies have evaluated the effects IUU fishing can have on the seafood industry.[46] Some documented impacts are described below.

Economic Loss. Researchers estimate that losses from IUU fishing are between $10 billion and $24 billion annually, representing 11-26 million tons of fish each year.[47] Furthermore, researchers estimated that IUU catches

---

[43] Jade Lindley, "Food Security Amidst Crime: Harm of Illegal Fishing and Fish Fraud on Sustainable Oceans," in *The Palgrave Handbook of Climate Resilient Societies*, ed. Robert C. Brears (Cham: Springer Nature Switzerland, 2022), pp. 733-751.

[44] In September 2024, some Members of Congress sent letters to the Council on Environmental Quality, NOAA, U.S. Department of State, U.S. Department of Labor, U.S. Department of Agriculture, U.S. Customs and Border Protection, U.S. Food and Drug Administration, and Federal Trade Commission. U.S. Congress, Natural Resources Committee Democrats, "Grijalva Leads Sweeping Bipartisan Request to Eight Federal Agencies to Examine Efforts to Combat IUU Fishing," September 4, 2024, https://democrats-naturalresources.house.gov/media/press-releases/grijalva-leads-sweeping-bipartisan-request-to-eight-federal-agencies-to-examine-efforts-to-combat-iuu-fishing.

[45] U.S. International Trade Commission (USITC), *Seafood Obtained via Illegal, Unreported, and Unregulated Fishing: U.S. Imports and Economic Impact on U.S. Commercial Fisheries*, Publication Number: 5168. Investigation Number 332-575, Washington, DC, February 2021, pp. 1-464. Hereinafter USITC, *Seafood Obtained via Illegal, Unreported, and Unregulated Fishing*.

[46] See, for example, Dana D. Miller and Ussif Rashid Sumaila, "Chapter 4: IUU Fishing and Impact on the Seafood Industry," in *Seafood Authenticity and Traceability*, eds. Amanda M. Naaum and Robert H. Hanner (Cambridge: Elsevier, 2016), pp. 83-95 (hereinafter Miller and Sumaila, "IUU Fishing and Impact"); Garnchanok Wongrak et al., "The Impact of the EU IUU Regulation on the Sustainability of the Thai Fishing Industry," *Sustainability*, vol. 13, no. 12 (2021), 6814, pp. 1-16 (hereinafter Wongrak et al., "Impact of the EU IUU Regulation"); and Don Liddick, "The Dimensions of a Transnational Crime Problem: The Case of IUU Fishing," *Trends in Organized Crime*, vol. 17 (2014), pp. 290-312 (hereinafter Liddick, "Dimensions of a Transnational Crime Problem").

[47] David J. Agnew, "Estimating the Worldwide Extent of Illegal Fishing," *PLoS One*, vol. 4, no. 2 (2009), e4570, pp. 1- 8.

represented 20%-32% of the wild-caught seafood imported into the United States by weight in 2011, with that value ranging from 11% to 13% in more recent estimates.[48] Fishers and shore-based businesses—such as processors, dealers, and vendors—may be harmed by the decrease in supplies of fish remaining available for legitimate harvest.

Threats to Economic Security. Experts have noted that IUU fishing can affect the stability and security of fishing activities at sea by impacting the economic value of fisheries resources.[49]

According to the National Intelligence Council, IUU fishing, along with surging worldwide demand for seafood and declining ocean health, poses an existential threat to global fisheries.[50]

Estimates of the global scale of IUU fishing are difficult to quantify in financial terms, but NOAA states that "there is little disagreement that it is in the billions, or even tens of billions, of dollars each year."[51] A 2020 study estimated that gross revenues associated with unreported fish catches worldwide potentially generate $9 billion to $17 billion in illicit proceeds annually.[52] The study further estimated corresponding economic impact losses due to the diversion of fish from the legitimate trade market as costing $26 billion to $50 billion annually and potential losses to countries' tax revenues worth approximately $2 billion to $4 billion annually.[53]

Sustainability and Market Value. IUU fishing can affect the sustainability of a fishery, primarily through overfishing. Overfishing can lead to reduced fishing opportunities for legitimate fishers, potential increased fishing costs (e.g., increased fuel costs from longer fishing trips to harvest fishes in deeper waters where fishes may be more abundant, or over a broader area), and lower profits, all of which can affect seafood market value and local economies.[54] For example, studies have estimated that the large concentration of IUU fishing off West Africa costs the region nearly $2 billion per year through the

---

[48] Ganapathiraju Pramod et al., "Estimates of Illegal and Unreported Fish in Seafood Imports to the USA," *Marine Policy*, vol. 48 (2014), pp. 102-113; and USITC, *Seafood Obtained via Illegal, Unreported, and Unregulated Fishing.*

[49] See, for example, Richard Barnes and Mercedes Rosello, "Fisheries and Maritime Security: Understanding and Enhancing the Connection," in *Maritime Security and the Law of the Sea: Help or Hindrance?*, eds. Malcolm D. Evans and Sofia Galani (Northampton, MA: Edward Elgar Publishing, 2020), pp. 48-82.

[50] NIC, Global Implications of IUU Fishing, p. 5.

[51] NOAA, NMFS, "Understanding Illegal, Unreported, and Unregulated Fishing."

[52] Ussif Rashid Sumaila et al., "Illicit Trade in Marine Fish Catch and Its Effects on Ecosystems and People Worldwide," *Science Advances*, vol. 6, no. 9 (February 2020), pp. 1-7.

[53] Ibid.

[54] Miller and Sumaila, "IUU Fishing and Impact."

reduction of fish stocks and economic losses to the tourism sector, with "hard-hitting" effects to families whose income relies on the fishing industry.[55] Some experts estimate this area has the highest recorded levels of IUU fishing in the world, with IUU fishing representing up to 40% of the total fish catch for the region.[56] Additionally, IUU fishing is estimated to cost developing nations between $2 billion and $15 billion in annual economic losses.[57] Because most IUU catches are not brought onshore to the country from whose waters they were taken, losses in port dues, revenue, and to transport and processing sectors can occur.[58] IUU fishing also may facilitate the inclusion of lower-quality or less valuable products in some seafood markets, which can impact the overall marketability of certain regional fisheries.[59] For example, overfishing associated with IUU fishing may reduce populations of more lucrative fish species in a given area—typically higher-level consumers in the food chain—and lead to increased fishing effort on other ecologically important species throughout the food chain.[60] Furthermore, IUU fishing may deplete less lucrative stocks that are critical food sources for human populations and marine wildlife, also affecting the sustainability of those stocks and the ecosystems they inhabit.[61]

Industry Reputation and Seafood Trade. IUU fishing can influence the reputations of certain national seafood industries and impact global seafood trade. Seafood industries from particular nations may receive certifications from other nations regarding their compatibility with the certifying nation's IUU fishing and sustainability standards. For example, the United States, through NOAA, identifies nations and entities that engage in IUU fishing and

---

[55] The term "hard hitting" is used to describe these effects in Hunter F. Donovan, "The Role of Corporations in Solving the Illegal, Unregulated, and Unreported (IUU) Fishing Crisis," *Ocean and Coastal Law Journal*, vol. 28, no. 1 (2023), pp. 177-212 (hereinafter Donovan, "Role of Corporations"). Miller and Sumaila, "IUU Fishing and Impact; and Sjarief Widjaja et al., "Illegal, Unreported, and Unregulated Fishing and Associated Drivers," in *The Blue Compendium: From Knowledge to Action for a Sustainable Ocean Economy*, eds. Jane Lubchenco and Peter M. Haugan (Cham: Springer Nature Switzerland, 2023), pp. 553-591.

[56] E. Drury O'Neill et al., "Socioeconomic Dynamics of the Ghanaian Tuna Industry: A Value-Chain Approach to Understanding Aspects of Global Fisheries," *African Journal of Marine Science*, vol. 40, no. 3 (2018), pp. 303-313; and Alkay Doumbouya et al., "Assessing the Effectiveness of Monitoring Control and Surveillance of Illegal Fishing: The Case of West Africa," *Frontiers in Marine Science*, vol. 4, no. 50 (2017), pp. 1-10.

[57] Liddick, "Dimensions of a Transnational Crime Problem."

[58] Ibid.

[59] Alan Reilly, "Overview of Food Fraud in the Fisheries Sector," FAO, FAO Fisheries and Aquaculture Circular No. 1165, 2018, pp. 1-21.

[60] For example, see Timothy E. Essington et al., "Fishing Through Marine Food Webs," *Proceedings of the National Academy of Sciences*, vol. 103, no. 9 (2006), pp. 3171-3175.

[61] Liddick, "Dimensions of a Transnational Crime Problem."

associated fishing activities that negatively affect protected living marine resources, in accordance with the High Seas Driftnet Fishing Moratorium Protection Act (commonly known as the Moratorium Protection Act; Title VI of P.L. 104-43).[62] NOAA additionally certifies nations if their regulatory programs regarding IUU fishing and protected living marine resources are compatible with those of the United States. The United States may deny port privileges or prohibit certain seafood imports for nations that receive a negative certification.[63]

Product mislabeling and fraud associated with IUU fishing can pose health risks to consumers through potential exposure to allergens, toxins, or contaminants.[64] In some cases, seafood products from particular nations may be identified as higher risk for health concerns, or for IUU fishing, and some nations may refuse to import said seafood products.[65] For example, the European Union can issue import bans on a nation's seafood if that nation's fishery products are associated with IUU fishing.[66] Negative certifications and seafood import bans can result in significant costs and economic effects to national and multinational seafood industries and markets.[67]

Traceability and Enforcement Costs. IUU fishing may lead to additional costs through national efforts to enhance seafood traceability and law enforcement. For example, increased permitting, registration, and regulatory compliance to combat IUU fishing can impose direct costs on the seafood community, including to fishers, seafood importers, and consumers.[68] Studies suggest that supply chain transparency enhancement programs can result in long-term sustainability benefits but also come with costs that typically are

---

[62] NOAA, NMFS, "NOAA Engagement with Nations and Entities Under the Moratorium Protection Act," https://www.fisheries.noaa.gov/international/international-affairs/noaa-engagement-nations-and-entities-under- moratorium. Hereinafter NOAA, NMFS, "NOAA Engagement with Nations and Entities Under the Moratorium Protection Act." For more information, see the "Magnuson-Stevens Fishery Conservation and Management Reauthorization Act of 2006 and High Seas Driftnet Fishing Moratorium Protection Act" section of this report.
[63] Ibid.; 16 U.S.C. §§1826h-1826k.
[64] Miller and Sumaila, "IUU Fishing and Impact."
[65] Ibid.
[66] Dae Eui Kim and Song Soo Lim, "Economic Impacts of the European Union Carding System on Global Fish Trade," *Marine Policy*, vol. 165 (2024), 106208, pp. 1-7 (hereinafter Kim and Lim, "Economic Impacts of the European Union Carding System"); and Wongrak et al., "Impact of the EU IUU Regulation."
[67] Kim and Lim, "Economic Impacts of the European Union Carding System"; and Juan He, "Unilateral Trade Measures Against Illegal, Unreported, and Unregulated Fishing: Unlocking a Paradigm Change in Trade- Environmental Partnerships?," *Journal of World Trade*, vol. 53, no. 5 (2019), pp. 759-782.
[68] Liddick, "Dimensions of a Transnational Crime Problem."

borne by producers in lower-income countries.[69] The U.S. seafood community estimates it has spent over $50 million on regulatory and paperwork compliance to address IUU fishing concerns for species covered through the U.S. Seafood Import Monitoring Program and has raised concerns about additional expenses it might incur were the program to expand.[70] Additionally, U.S. importers found to violate national traceability and labeling requirements for imported seafood—or found to have not reported falsely labeled products—may be subject to penalties.[71] Some note that those in the seafood industry are motivated to voluntarily participate in traceability initiatives and that consumer demand for sustainable products can serve as a traceability motivator and could lead to expanded trade opportunities.[72] In some cases, stricter multinational policies regarding supply chain transparency may deter the entry of mislabeled seafood into the market and facilitate consumers' access to accurate information.[73]

Are Transnational Crimes Associated with IUU Fishing?

IUU fishing, as well as some seafood industries, may be associated with other illegal activities, such as human and other forms of trafficking (e.g., arms, drugs, wildlife), labor exploitation, and organized crime.[74] For example, some experts have reported at least 100 Russian trawlers were operating in "mafia-

---

[69] For example, John Virdin et al., "Combatting Illegal Fishing Through Transparency Initiatives: Lessons Learned from Comparative Analysis of Transparency Initiatives in Seafood, Apparel, Extractive, and Timber Supply Chains," *Marine Policy*, vol. 138 (2022), 104984, pp. 1-11. Hereinafter Virdin et al., "Combatting Illegal Fishing Through Transparency Initiatives."

[70] For more information, see the "What Is the Seafood Import Monitoring Program?" section of this report. Bhavana Scalia-Bruce, "NOAA Issues Fines for Importer Violating SIMP Regulations," *Seafood Source*, April 28, 2023, https://www.seafoodsource.com/news/supply-trade/noaa-issues-fines-for-importer-violating-simp-regulations.

[71] Ibid.

[72] Virdin et al., "Combatting Illegal Fishing Through Transparency Initiatives."

[73] Ibid.; Michaela Fox et al., "The Seafood Supply Chain from a Fraudulent Perspective," *Food Security*, vol. 10 (2018), pp. 939-963; and Miller and Sumaila, "IUU Fishing and Impact."

[74] FAO, *State of World Fisheries and Aquaculture*, 2024, p. 222; Gohar A. Petrossian et al., "Organized Crime in the Fisheries Sector," in *The Private Sector and Organized Crime*, eds. Yuliya Zabyelina and Kimberly L. Thachuk (New York: Routledge, 2023), pp. 132-148 (hereinafter Petrossian et al., "Organized Crime in the Fisheries Sector"); and Kaija Metuzals et al., "One Fish, Two Fish, IUU, and No Fish: Unreported Fishing Worldwide," in *Handbook of Marine Fisheries Conservation and Management*, eds. R. Quentin Grafton et al. (Oxford: Oxford University Press, 2009), pp. 166-180 (hereinafter Metuzals et al., "One Fish, Two Fish, IUU, and No Fish").

style gangs" during the mid-2000s and were associated with large-scale illegal and unreported fishing.[75] Some criminals may leave other illegal industries to engage in IUU fishing due to the perception that IUU fishing may be more lucrative and less dangerous.[76] Fishers may be vulnerable to recruitment by criminal organizations seeking to use vessels for illegal operations. Some experts infer that IUU fishing may be classified as organized crime based on the description in the UN Convention Against Transnational Organized Crime.[77] FAO and U.S. definitions of IUU fishing are specific to fisheries resource laws, whereas other types of illegal activities are subject to other national laws and international agreements.[78] In 2021, the Biden Administration directed federal departments and agencies to take actions within their respective authorities to enhance efforts to counter transnational organized crime, including organizations engaged in illegal fishing.[79]

Human Trafficking and Labor Exploitation. According to FAO, "migrant workers are particularly exposed to modern slavery, bondage, forced labour and other abuses, which have been associated with IUU fishing."[80] Several factors make the fishing sector susceptible to human trafficking.[81] Traffickers often recruit fishers living in impoverished areas or in areas with political instability by making false claims of high wages or immigration assistance.[82] In a report to Congress, the Departments of Commerce and State identified 29 countries or territories most at risk for human trafficking, including forced labor, in the seafood sector.[83] Traffickers charge fishers a recruitment fee to get them employment and then sell the fee obligation to a fishing vessel

---

[75] Metuzals et al., "One Fish, Two Fish, IUU, and No Fish."
[76] Liddick, "Dimensions of a Transnational Crime Problem."
[77] Petrossian et al., "Organized Crime in the Fisheries Sector"; and United Nations Office on Drugs and Crime, *United Nations Convention Against Transnational Organized Crime and the Protocols Thereto*, New York, 2004, pp. 1-82.
[78] Refer to the "What Is IUU Fishing?" section of this report.
[79] The White House, "Executive Order on Establishing the United States Council on Transnational Organized Crime," December 15, 2021, https://www.whitehouse.gov/briefing-room/presidential-actions/2021/12/15/executive-order-on-establishing-the-united-states-council-on-transnational-organized-crime/.
[80] Ibid., p. 171.
[81] Departments of Commerce and State, *Human Trafficking in the Seafood Supply Chain*.
[82] For example, see Department of Justice, *Report on Human Trafficking in Fishing in International Waters*, Report to Congress, January 2021, pp. 1-51, see pp. 7-8.
[83] The 29 countries and territories identified as most at risk for human trafficking were Bangladesh, Burma, Cambodia, Cameroon, Ecuador, Fiji, Gabon, Ghana, Guinea, Honduras, Indonesia, Ireland, Kenya, Madagascar, Mauritania, North Korea, Pakistan, Papua New Guinea, the People's Republic of China, Philippines, Seychelles, Sierra Leone, South Africa, South Korea, Taiwan, Tanzania, Thailand, Vanuatu, and Vietnam. Departments of Commerce and State, *Human Trafficking in the Seafood Supply Chain*.

captain.[84] Unable to pay off their recruitment fee to the vessel captain, fishers are forced to remain with the vessel or firm, a situation referred to as *debt bondage*.

Fishers on DWF vessels are inherently isolated on the high seas.[85] The Outlaw Ocean Project, a nonprofit journalism organization, reports that some DWF vessels remain at sea for years, relying on refrigeration vessels to transport fish catch from the vessels back to shore.[86] Under these isolated conditions, fishers are unable to report abuse or escape.[87] Furthermore, migrants on fishing vessels may be unable to communicate with operators due to language barriers.[88]

IUU fishing operations may violate basic safety standards and deny crew members' fundamental rights, such as agreed-on terms and conditions of their labor. An investigation by the Outlaw Ocean Project documented several possible abuses on at least 119 fishing vessels since 2013, including debt bondage, wage withholding, excessive working hours, beatings, passport confiscation, lack of timely access to medical care, and deaths from neglect or violence.[89] Forced labor is not limited to fishing vessels and may extend to the seafood processing sector.

The U.S. definition for IUU fishing does not include human trafficking, forced labor, and other related crimes (refer to "What Is IUU Fishing?"). Congress may consider directing the Secretary of Commerce, through NOAA NMFS, to provide a new definition for IUU fishing that includes human trafficking, forced labor, and other related crimes. Some in Congress have called NOAA's definition "narrow" and argue that updating the definition would align with international standards.[90] Others may contend that additional

---

[84] For example, see Ian Urbina, "Lawless Ocean: The Link Between Human Rights Abuses and Overfishing," Outlaw Ocean Project, November 20, 2019, https://www.theoutlawocean.com/reporting/the-link-between-human-rights-abuses-and-overfishing/; and Departments of Commerce and State, *Human Trafficking in the Seafood Supply Chain*.

[85] Department of Justice, *Report on Human Trafficking in Fishing in International Waters*, Report to Congress, January 2021, p. 5, https://www.justice.gov/crt/page/file/1360366/dl; and Departments of Commerce and State, *Human Trafficking in the Seafood Supply Chain*.

[86] An Outlaw Ocean Project investigation found that "foreign and Chinese workers on these fishing ships stay at sea for more than three years." See Outlaw Ocean Project, "China: The Superpower of Seafood," https://www.theoutlawocean.com/investigations/china-the-superpower-of-seafood/findings/.

[87] Departments of Commerce and State, *Human Trafficking in the Seafood Supply Chain*.

[88] Ibid.

[89] Outlaw Ocean Project, "China: The Superpower of Seafood," https://www.theoutlawocean.com/investigations/china-the-superpower-of-seafood/findings/.

[90] Letter from The Honorable Raúl M. Grijalva, Member of Congress, The Honorable Jared Huffman, Member of Congress, and The Honorable Frank Pallone Jr., Member of

resources may be needed to account for these considerations through new or existing programs to address IUU fishing.

Arms, Drugs, and Wildlife Trafficking. Arms, drugs, and wildlife trafficking crimes often have no direct connection with fishing operations but take place on fishing vessels, "using the fishing operation as a cover, opportunity or means to commit such crimes."[91] Some experts contend that human trafficking and forced labor should not be considered in the same context as arms, drugs, and wildlife trafficking because this conflation "runs the risk of criminalizing victims of modern slavery or forced labor" by potentially associating them with these practices.[92]

What International Agreements Address IUU Fishing?

FAO provides an international framework to address IUU fishing globally and implements several international fisheries legal instruments. Other multilateral agreements outside of the FAO framework also have been established to address IUU fishing. Selected multilateral fisheries instruments to which the United States has agreed, both internal and external to the FAO framework, are described in chronological order below.

1995 UN Fish Stocks Agreement

The 1995 UN Fish Stocks Agreement elaborates on the UNCLOS principle that nations should cooperate to ensure the long-term conservation of fisheries resources and to promote optimum utilization of these resources.[93] In general, the provisions of the 1995 UN Fish Stocks Agreement include conservation and management measures (Articles 5-7); mechanisms for international cooperation, such as RFMOs (Articles 8-16); duties of flag nations (Article 18); compliance and enforcement (Articles 19-23); and dispute settlement (Articles 27-32).

Congress, et al. to The Honorable Joseph R. Biden Jr., President of the United States, March 11, 2024, https://democrats-naturalresources.house.gov/imo/media/doc/2024-03-11_moc_letter_to_president_biden_re_iuu_fishing.pdf.
[91] FAO, "Links Between IUU Fishing and Crimes in the Fisheries Sector," https://www.fao.org/iuu-fishing/ background/links-crimes/en/.
[92] Mary Mackay et al., "The Intersection Between Illegal Fishing, Crimes at Sea, and Social Well-Being," *Frontiers in Marine Science*, vol. 7 (2020), pp. 1-9, see p. 7.
[93] United Nations, "Fish Stocks Agreement."

The 1995 UN Fish Stocks Agreement mostly applies to areas beyond the limits of the EEZ and establishes a framework for RFMOs to manage and conserve fish stocks in certain high seas areas. The agreement requires parties to have their commercial fishing vessels accurately collect and share fisheries data. These efforts can assist with assessing and addressing the ecological and economic effects from IUU fishing. The United States is a party to the 1995 UN Fish Stocks Agreement.[94]

Port State Measures Agreement

The Agreement on Port State Measures to Prevent, Deter, and Eliminate Illegal, Unreported, and Unregulated Fishing (commonly known as the Port State Measures Agreement, or PSMA) entered into force in June 2016 and is recognized as the first binding international agreement to target IUU fishing.[95] The PSMA aims to prevent, deter, and eliminate IUU fishing by preventing vessels participating in IUU fishing activities from using ports and bringing their catches onshore.[96] According to FAO, "the provisions of the PSMA apply to fishing vessels seeking entry into a designated port of a State which is different to their flag State."[97] Port state measures focus on vessel inspections, which may limit transport of illegally harvested products through certain ports. These measures also may be a disincentive to engaging in illegal activity, because they can make transshipments of fish and the resupply of fishing vessels more costly. The PSMA additionally serves as a basis for the FAO Global Record of Fishing Vessels, Refrigerated Transport Vessels, and Supply Vessels, which is an online repository used to help deter and eliminate IUU fishing activities, for use by inspectors, administrators, managers, and

---

[94] The United States and 25 other nations signed the agreement on December 4, 1995, the first day it was open for signature, and the 1995 UN Fish Stocks Agreement entered into force upon ratification of the 30th nation on December 11, 2001. The U.S. Senate agreed to a resolution of advice and consent to ratification of this agreement on June 27, 1996 (U.S. Congress, Senate, Agreement for the Implementation of the United Nations Convention of the Law of the Sea of 10 December 1982 Relating to Fish Stocks, 104th Cong., 2nd sess., February 1996, Treaty Doc. 104-24 (Washington, DC: GPO, 1996)). China is not a party to the 1995 UN Fish Stocks Agreement. United Nations, "Chronological Lists of Ratifications of, Accessions and Successions to the Convention and the Related Agreements," updated October 24, 2023, https://www.un.org/depts/los/reference_files/chronological_lists_of_ratifications.htm.

[95] FAO, Port State Measures Agreement.

[96] FAO, "Illegal, Unreported, and Unregulated (IUU) Fishing: Agreement on Port State Measures (PSMA)," https://www.fao.org/iuu-fishing/international-framework/psma/en/.

[97] Ibid.

other stakeholders.⁹⁸ The PSMA has 79 signatories, including the United States.⁹⁹

Agreement to Prevent Unregulated High Seas Fisheries in the Central Arctic Ocean

On July 25, 2021, the Agreement to Prevent Unregulated High Seas Fisheries in the Central Arctic Ocean entered into force. Canada, China, Denmark (in respect to the Faroe Islands and Greenland), the European Union, Iceland, Japan, Norway, Russia, South Korea, and the United States are signatories to the agreement. The agreement aims to prevent unregulated fishing in the high seas portion of the central Arctic Ocean and facilitate joint scientific research and monitoring.¹⁰⁰ Signatories agreed to a 16-year moratorium on commercial fishing in the central Arctic Ocean.¹⁰¹ The moratorium is in place until at least 2037.

World Trade Organization Agreement on Fisheries Subsidies[102]

On June 17, 2022, WTO members finalized an agreement aimed at curbing fisheries subsidies.¹⁰³ The 2022 WTO Agreement on Fisheries Subsidies

---

⁹⁸ FAO, "Global Record of Fishing Vessels, Refrigerated Transport Vessels, and Supply Vessels," https://www.fao.org/ global-record/en/.

⁹⁹ The U.S. Senate gave its advice and consent to ratification of the PSMA in 2014 (U.S. Congress, Senate, *Agreement on Port State Measures to Prevent, Deter, and Eliminate Illegal, Unreported, and Unregulated Fishing, Done at the Food and Agriculture Organization of the United Nations*, Rome, Italy, November 22, 2009, 113th Cong., 2nd sess., April 2014, Treaty Doc. 112-4 (Washington, DC: GPO, 2014). China is not a signatory to the PSMA. FAO, "Parties to the PSMA," https://www.fao.org/port-state-measures/background/parties-psma/en/.

¹⁰⁰ U.S. Department of State, "The Agreement to Prevent Unregulated High Seas Fisheries in the Central Arctic Ocean Enters into Force," June 25, 2021, https://www.state.gov/the-agreement-to-prevent-unregulated-high-seas-fisheries-in-the-central-arctic-ocean-enters-into-force/.

¹⁰¹ The Arctic Council, "An Introduction to the International Agreement to Prevent Unregulated Fishing in the High Seas of the Central Arctic Ocean," June 25, 2021, https://arctic-council.org/news/introduction-to-international-agreement-to-prevent-unregulated-fishing-in-the-high-seas-of-the-central-arctic-ocean/.

¹⁰² Liana Wong, Analyst in International Trade and Finance, contributed to this section on the World Trade Organization (WTO) Agreement on Fisheries Subsidies. For more information about this agreement, see CRS In Focus IF11929, *World Trade Organization Fisheries Subsidies Negotiations*, by Liana Wong.

¹⁰³ WTO, "Agreement on Fisheries Subsidies," https://www.wto.org/english/tratop_e/rulesneg_e/fish_e/fish_e.htm.

prohibits governments from providing subsidies to fisheries participating in IUU fishing and fishing of already overfished stocks.[104] It does not address other key issues laid out in a 2017 WTO Ministerial Conference mandate, such as subsidies contributing to overcapacity and special and differential treatment for developing country members, which would include China.[105] The 2022 agreement includes a sunset provision and is to automatically terminate if members fail to agree on "comprehensive disciplines" within four years after entry into force. As of September 2024, 81 WTO members, including the United States, have formally accepted the agreement.[106] The agreement is to enter into force after two- thirds of WTO members ratify it.[107] WTO members continue to negotiate on outstanding issues that were not addressed in the 2022 agreement, including subsidies that contribute to overcapacity, exceptions for certain subsidies, and special and differential treatment for developing countries.[108]

How Can Regional Fisheries Management Organizations Deter IUU Fishing?

RFMOs are international fishery management bodies established to conserve and manage transboundary fish stocks (i.e., fish that move across maritime zones), such as tuna or other highly migratory species.[109] For example, the

---

[104] WTO, *Agreement on Fisheries Subsidies*, WT/MIN(22)/33, June 22, 2022.

[105] China, which self-designated as a developing country, has indicated that it would not take advantage of the flexibilities made available to developing countries under special and differential treatment provisions in a potential expansion of the WTO Agreement on Fisheries Subsidies. See *South China Morning Post*, "China Won't 'Compete' with Other Developing Nations During WTO Fishing Talks, Subsidies Set to be Discussed in Abu Dhabi," February 16, 2024; and WTO, *Fisheries Subsidies Ministerial Decision of 13 December 2017*, WT/MIN(17)/64, December 18, 2017.

[106] Both China and Russia have accepted the agreement. WTO, "Members Submitting Acceptance of Agreement on Fisheries Subsidies," https://www.wto.org/english/tratop_e/rulesneg_e/fish_e/fish_acceptances_e.htm.

[107] 110 WTO members need to accept the agreement for it to enter into force. WTO, "Jordan Formally Accepts Agreement on Fisheries Subsidies," July 23, 2024, https://www.wto.org/english/news_e/ news24_ e/fish_23jul24_e.htm.

[108] WTO, "DDG Ellard, at FAO Meeting, Urges Completion of Critical Work on Fisheries Subsidies," July 10, 2024, https://www.wto.org/english/news_e/news24_e/ddgae_10jul 24_e.htm.

[109] Estimates for the number of RFMOs globally can vary because there is no single definition for how an international body qualifies as an RFMO. FAO, *Regional Fisheries Management Organizations and Advisory Bodies: Activities and Developments, 2000-2017*, FAO Fisheries and Aquaculture Technical Paper 651, 2020.

International Commission for the Conservation of Atlantic Tunas is one of five RFMOs focused on the management of tuna and tuna-like species (also known as tuna RFMOs).[110] Other RFMOs manage fishing activities occurring within a specific region of the high seas; for example, the South Pacific Regional Fisheries Management Organization manages high seas fisheries in the southern Pacific (ranging east-west from South America to Australia).[111] RFMO membership is open to nations with an interest in fishery resources within a given region.[112]

Fisheries management and enforcement vary between RFMOs, which can have implications for regional and global efforts to curb IUU fishing. For example, one way to deter IUU fishing within an RFMO is through high seas boarding and inspection (HSBI) of suspect vessels. Not all RFMOs have adopted a HSBI regime, though.[113] As another example, some RFMOs, such as the Western and Central Pacific Fisheries Commission,[114] have an observer program that requires observers to be stationed on fishing vessels to collect biological data and to monitor compliance of fishing observations.[115] Enforcement agents may use information collected by at-sea observers to investigate and prosecute violations. However, at-sea observers may be subject to intimidation and harassment, especially on vessels conducting illegal activities.[116] Other RFMO actions, including in-depth collaborations with stakeholders, required use of a catch documentation scheme, and maintenance of an "IUU fishing vessel blacklist," may help curb IUU fishing.

---

[110] International Commission for the Conservation of Atlantic Tunas, *Basic Texts*, 7th Revision, Madrid, 2019; and Tuna-org, "Tuna-org," https://www.tuna-org.org/index.htm.

[111] South Pacific Regional Fishery Management Organization, https://www.sprfmo.int/.

[112] European Commission, "Regional Fisheries Management Organisations (RFMOs)," https://oceans-and-fisheries.ec.europa.eu/fisheries/international-agreements/regional-fisheries-management-organisations-rfmos_en; and Pew Charitable Trusts, "FAQ: What Is a Regional Fishery Management Organization?," https://www.pewtrusts.org/en/research-and-analysis/fact-sheets/2012/02/23/faq-what-is-a-regional-fishery-management-organization.

[113] For example, FAO, *High Seas Boarding and Inspection of Fishing Vessels: Discussion of Goals, Comparison of Existing Schemes and Draft Language*, September 2003.

[114] Western & Central Pacific Fisheries Commission, "Regional Observer Programme," https://www.wcpfc.int/regional-observer-programme.

[115] Fishing observers also exist at the national level. Foreign fishing vessels may be required to have an at-sea observer to operate in the EEZs of certain coastal nations, including the United States, Australia, New Zealand, the Philippines, and Portugal. François Mosnier et al., *Bonding with Observers*, Planet Tracker, April 2021, p. 3.

[116] Some of the dangers facing at-sea observers are the same as those faced by fishing crew members. Some at-sea observers have gone missing while working on fishing vessels. For example, Human Rights at Sea, *Investigative Report and Case Study Fisheries Abuses and Related Deaths at Sea in the Pacific Region*, December 2017, p. 8.

For example, the Commission for the Conservation of Antarctic Marine Living Resources implemented some of these actions regarding IUU fishing and saw a greater than 90% reduction of IUU fishing for Patagonian toothfish (Dissostichus eleginoides; also known as Chilean sea bass) in the Antarctic region.[117]

The United States belongs to nine RFMOs.[118] Most of these RFMOs have an HSBI regime (Table 1). Congress has recognized the potential for HSBI to help deter IUU fishing on the high seas. In 2019, Congress directed the Secretary of State, in consultation with the Secretary of Commerce, to coordinate with RFMOs, along with other international organizations, to "enhance regional responses to IUU fishing and related transnational organized illegal activities."[119]

Not all high seas areas have RFMOs or are covered by an RFMO with management and enforcement mandates to counter IUU fishing.[120] High seas areas in the South China Sea, central Arctic Ocean, southwest Atlantic, and off the Horn of Africa do not have geographically specific RFMOs. The patchwork management and enforcement of high seas fisheries may contribute to unabated IUU fishing. According to some experts, the establishment of new RFMOs could reduce the number of fisheries conflicts between neighboring nations (e.g., in the South China Sea).[121] Congress may consider whether more resources and greater diplomatic support could help in the coordination of fishery management in adjacent territorial waters or in regions currently without RFMOs.

---

[117] Miller and Sumaila, "IUU Fishing and Impact"; Henrik Österblom et al., "Reducing Illegal Fishing in the Southern Ocean: A Global Effort," *Solutions*, vol. 4 (2015), pp. 72-79; and Commission for the Conservation of Antarctic Marine Living Resources, "Illegal, Unreported, and Unregulated (IUU) Fishing," https://www.ccamlr.org/en/comp liance/illegal-unreported-and-unregulated-iuu-fishing.

[118] NOAA, "International and Regional Fisheries Management Organizations," https://www.fisheries.noaa.gov/ international-affairs/international-and-regional-fisheries-management-organizations; and U.S. Department of State, "International Fisheries Management," https://www.state.gov/key-topics-office-of-marine-conservation/inter natio nal- fisheries-management/.

[119] Section 3541 of the Maritime Security and Fisheries Enforcement Act (Division C, Title XXXV, Subtitle C, of P.L. 116-92; commonly known as the Maritime SAFE Act); 16 U.S.C. §8011.

[120] For example, see FAO, "Marine Protected Areas in the High Seas," https://www.fao.org/fishery/en/topic/16204.

[121] For example, see Shui-Kai Chang et al., "A Step Forward to the Joint Management of the South China Sea Fisheries Resources: Joint Works on Catch, Management Measures, and Conservation Issues," *Marine Policy*, vol. 116 (2020), pp. 1-13.

Table 1. Regional fisheries management organizations (with U.S. membership, by alphabetical order)

Regional Fisheries Management Organization	Ocean Basin	Purpose (Species/Regional)	High Seas Boarding and Inspection	Number of Members	Membership of Top Global Marine Capture Fisheries[a]
Commission for the Conservation of Antarctic Marine Living Resources (CCAMLR)	Southern Ocean	Regional	Yes	27	Chile, China, India, Japan, Norway, Peru, Russia, South Korea, United States
Inter-American Tropical Tuna Commission (IATTC)	Eastern Pacific Ocean (Canada to Chile)	Species (tuna and tuna-like species)	Yes	21	China, Japan, Peru, South Korea, United States
International Commission for the Conservation of Atlantic Tunas (ICCAT)	Atlantic Ocean	Species (tuna and tuna-like species)	Yes	52	China, Japan, Norway, Russia, United States
North Atlantic Salmon Conservation Organization (NASCO)	North Atlantic Ocean	Species (Atlantic salmon)	No	7	Norway, Russia, United States
North Pacific Anadromous Fish Commission (NPAFC)	North Pacific Ocean	Species (Pacific salmon and steelhead trout)	Yes	5	Japan, Russia, South Korea, United States
North Pacific Fisheries Commission (NPFC)	North Pacific Ocean	Regional	Yes	9	China, Japan, Russia, South Korea, United States
Northwest Atlantic Fisheries Organization (NAFO)	Northwest Atlantic Ocean	Regional	Yes	13	Japan, Norway, Russia, South Korea, United States
South Pacific Regional Fisheries Management	South Pacific Ocean	Regional	Yes	16	Chile, China, Peru, Russia, South Korea, United States

Table 1. (Continued)

Regional Fisheries Management Organization	Ocean Basin	Purpose (Species/Regional)	High Seas Boarding and Inspection	Number of Members	Membership of Top Global Marine Capture Fisheries[a]
Organization (SPRFMO)					
Western and Central Pacific Fisheries Commission (WCPFC)	Western and Central Pacific Ocean	Species (tuna and tuna-like species)	Yes	26	China, Indonesia, Japan, United States[b]

Sources: CCAMLR, https://www.ccamlr.org/en; CCAMLR, "System of Inspection," https://www.ccamlr.org/en/ compliance/inspections; IATTC, https://www.iattc.org/; IATTC, Resolution on Boarding and Inspection Procedures, Inter-American Tropical Tuna Commission 90th Meeting, June 27-July 1, 2016, https://www.iattc.org/ GetAttachment/ecf7172a-57ea-4c20-811c-f1cdafaf8394/IATTC-90-PROP-H-1_REV1-USA-Boarding-and-Inspection-Procedures-track-changes.pdf; ICCAT, https://www.iccat.int/en/; ICCAT, "ICCAT Joint Scheme of International Inspection," https://www.iccat.int/en/Inspection.html; NAFO, https://www.nafo.int/; Jean-Jacques Maguire et al., Report of the Third NASCO Performance Review, NASCO, Final Report, CNL(23)17rev, March 14, 2023, p. 65, https://nasco.int/wp-content/uploads/2023/05/CNL2317rev_Report-of-the-Third-NASCO-Performance-Review.pdf; NAFO, Conservation and Enforcement Measures 2024, NAFO/COM Doc. 24-01, 2024, https://www.nafo.int/Portals/0/PDFs/COM/2024/comdoc24-01.pdf; NASCO, https://nasco.int/; National Oceanic and Atmospheric Administration (NOAA), "International and Regional Fisheries Management Organizations," https://www.fisheries.noaa.gov/international-affairs/international-and-regional-fisheries-management-organizations; NPAFC, https://www.npafc.org/; NPAFC, "Frequently Asked Questions," https://www.npafc.org/ faq/#iuu; NPFC, https://www.npfc.int/; NPFC, "NPFC High Seas Boarding & Inspection," https://www.npfc.int/npfc-high-seas-boarding-inspection; SPRFMO, https://www.sprfmo.int/; SPRFMO, Conservation and Management Measure for High Seas Boarding and Inspection Procedures for the South Pacific Regional Fisheries Management Organization (supersedes CMM 11-2015), CMM 11-2023, https://www.sprfmo.int/assets/Fisheries/Conservation-and-Management-Measures/2023-CMMs/CMM-11-2023-Boarding-and-Inspection_29Mar23.pdf; U.S. Department of State, "International Fisheries Management," https://www.state.gov/key-topics-office-of-marine-conservation/international-fisheries-management/; WCPFC, https://www.wcpfc.int/home; and WCPFC, "High Seas Boarding & Inspection," https://www.wcpfc.int/high-seas-boarding-inspection.

Notes: FAO = Food and Agriculture Organization of the United Nations. The United States participates as an observer in the Commission for the Conservation of Southern Bluefin Tuna, Indian Ocean Tuna Commission, and Southern Indian Ocean Fisheries Agreement. According to NOAA, the United States signed the Convention for the South East Atlantic Fisheries Organization but has not ratified it because there is no U.S. fishing in the convention area at present. Marine capture fisheries are those in which fishery species are directly harvested from marine waters.

[a] The top 10 marine capture fisheries producers in 2022 by nation as identified by FAO are China (14.8%), Indonesia (8.6%), Peru (6.6%), Russia (5.9%), United States (5.3%), India (4.5%), Vietnam (4.3%), Japan (3.6%), Norway (3.1%), and Chile (2.8%). FAO, The State of World Fisheries and Aquaculture: Blue Transformation in Action, 2024, see Table 6 on p. 29.

[b] Vietnam is listed as a cooperating non-member of the WCPFC.

What U.S. Laws Address IUU Fishing?

Congress has passed several laws aimed at directly or indirectly addressing IUU fishing activities occurring within waters under U.S. jurisdiction and/or the high seas. Some of these laws address the impacts of marine biodiversity loss associated with IUU fishing; others address the law enforcement aspects of IUU fishing. Selected laws are discussed in chronological order below. Additionally, Congress has included directives in appropriations acts regarding agency programs and activities to address IUU fishing.

Magnuson-Stevens Fishery Conservation and Management Reauthorization Act of 2006 and High Seas Driftnet Fishing Moratorium Protection Act

In 2006, Congress amended the Moratorium Protection Act to include considerations for IUU fishing.[122] These amendments were included as part of the Magnuson-Stevens Fishery Conservation and Management Reauthorization Act of 2006 (MSRA).[123] Through the MSRA, Congress also amended Section 2(a) of the Magnuson-Stevens Fishery Conservation and Management Act (P.L. 94-265) to add the finding that "international cooperation is necessary to address IUU fishing and other fishing practices that may harm the sustainability of living marine resources and the U.S. fishing industry."[124]

The Moratorium Protection Act originally was enacted to build on legislation that controls and prohibits large-scale driftnet fishing within the U.S. EEZ and on the high seas.[125] It also was enacted to prohibit the United States from entering into any international agreement regarding living marine resource conservation and management that would prevent full implementation of the global moratorium on large-scale driftnet fishing on the

---

[122] For more on the Moratorium Protection Act, also see the "How Does IUU Fishing Impact the Seafood Industry?" section of this report.

[123] 16 U.S.C. §§1826h-1826k, 1829.

[124] 16 U.S.C. §1801(a)(12).

[125] Other laws on which the act builds include the Driftnet Impact Monitoring, Assessment, and Control Act of 1987 (Title IV of P.L. 100-220); the Driftnet Act Amendments of 1990 (P.L. 101-627); and the High Seas Driftnet Fisheries Enforcement Act (Title I of P.L. 102-582).

high seas.[126] Congress included specific provisions in the MSRA authorizing the Secretary of Commerce to share information on multinational harvesting and processing capacity and IUU fishing in U.S. waters, the high seas, and areas covered by international fishery management agreements with foreign law enforcement and international organizations.[127] Congress also authorized the Secretary of Commerce to enhance enforcement and technological capabilities to locate and identify IUU fishing vessels on the high seas and encroachments of foreign fishing vessels into the U.S. EEZ.[128]

Congress further required the Secretary of Commerce to produce a biennial report identifying nations whose vessels have participated in IUU fishing and in fishing practices that lead to unregulated bycatch of protected species and sharks, among other unsustainable fishing practices, on the high seas or in any nation's EEZ.[129] The Secretary of Commerce is to identify and list in the report any nations that have not "adopted, implemented, and enforced" a regulatory program governing those activities comparable in effectiveness to that of the United States.[130] The report also is to identify nations and entities with which the United States will work over a two-year period to address IUU fishing, among other actions.[131]

NOAA has submitted these reports to Congress since 2009. The reports include certification determinations on whether identified nations took actions to remedy identified IUU and unsustainable fishing activities.[132] In the event of a negative certification, NOAA may deny U.S. port access to fishing vessels of that nation and may impose import restrictions on its fish or fish products.[133] For example, in October 2024, NOAA is to deny U.S. port entry to fishing

---

[126] For example, as expressed in Resolution 46/215 of the UN General Assembly. UN General Assembly, *46/215. Large-Scale Pelagic Drift-Net Fishing and Its Impact on the Living Marine Resources of the World's Oceans and Seas*, 46th Session, 79th Plenary Meeting, December 20, 1991, pp. 147-148.

[127] 16 U.S.C. §1829(b)(1).

[128] 16 U.S.C. §1829(b).

[129] 16 U.S.C. §§1826h, 1826j-1826k.

[130] 16 U.S.C. §§1826h, 1826k. These provisions also include the identification of nations that have not adopted conservation measures comparable to those of the United States "to provide for the conservation of sharks, including measures to prohibit removal of any of the fins of a shark, including the tail, before landing the shark in port," as amended through the Shark Conservation Act of 2010 (P.L. 111-348). 16 U.S.C. §1826k(a)(1)(B).

[131] 16 U.S.C. §1826h; NOAA, NMFS, "NOAA Engagement with Nations and Entities Under the Moratorium Protection Act."

[132] Ibid.

[133] 16 U.S.C. §§1826a, 1826j(d)(3), 1826k(c)(5). As also included in amendments to the High Seas Driftnet Fishing Moratorium Protection Act through the Illegal, Unreported, and Unregulated Fishing Enforcement Act of 2015 (P.L. 114-81).

vessels from 17 nations, including China, Mexico, and Russia, as a result of negative certification under the Moratorium Protection Act.[134] Some stakeholders contend that the U.S. denial of port entry to vessels from these nations is unlikely to influence their fishing behaviors because they generally stay away from U.S. ports.[135]

Together with these reports, the Secretary of Commerce, in consultation with the Secretary of State and other relevant parties, is to take actions to improve the effectiveness of international fishery management and conservation actions by urging international fishery management organizations (e.g., RFMOs) of which the United States is a member to address IUU fishing.[136]

Actions include market-related measures, vessel identification lists and monitoring systems, port state controls (i.e., prohibiting vessel port access), and import prohibitions, among other measures to prevent IUU fishing.[137] As discussed above (see "What is IUU Fishing?"), the Moratorium Protection Act also required the Secretary of Commerce to publish a definition for IUU fishing.[138]

Maritime Security and Fisheries Enforcement Act

In 2019, Congress passed the Maritime Security and Fisheries Enforcement Act (commonly known as the Maritime SAFE Act; Division C, Title XXXV, Subtitle C, of P.L. 116-92) as part of the National Defense Authorization Act for Fiscal Year 2020.[139] The Maritime SAFE Act seeks to support a whole-of-government approach to counter IUU fishing, improve data sharing, support efforts to counter IUU fishing in priority regions around the world, increase global transparency and traceability across the seafood chain, improve global

---

[134] NOAA denied U.S. port entry to vessels from Algeria, Barbados, China, Côte d'Ivoire, Cyprus, France, Greece, Italy, Malta, Mexico, Namibia, Russia, Senegal, Spain, Trinidad and Tobago, Tunisia, and Turkey. NOAA, "NOAA Fisheries Denies U.S. Port Privileges to Certain Fishing Vessels from 17 Nations," September 10, 2024, https://www.fisheries.noaa.gov/feature-story/noaa-fisheries-denies-us-port-privileges-certain-fishing-vessels-17- nations.

[135] Daniel Cusick, "NOAA Puts 17 Nations on Notice Over Illegal Fishing," E&E News, September 11, 2024, https://subscriber.politicopro.com/article/eenews/2024/09/11/noaa-puts-17-nations-on-notice-over-illegal-fishing- 00178646.

[136] 16 U.S.C. §1826i(a).

[137] Ibid.

[138] 16 U.S.C. §1826j(e).

[139] 16 U.S.C. §§8001 et seq.

enforcement operations against IUU fishing, and prevent the use of IUU fishing as a financing source for transnational crime.[140] The Maritime SAFE Act also established the Interagency Working Group on IUU Fishing to support and coordinate a government-wide "approach to counter IUU fishing and related threats to maritime security" globally.[141]

The Maritime SAFE Act also directs certain agencies to carry out specific activities to address IUU fishing. Some examples of these activities are listed below.

- The Department of State, in consultation with the Secretary of Commerce (i.e., through NOAA, as in the case for the below examples), shall coordinate with RFMOs, FAO, and other relevant international organizations to enhance regional responses to IUU fishing and transnational organized illegal activities.[142]
- The Department of State may engage its chiefs of mission in relevant countries to examine IUU fishing.[143]
- The Department of State, in consultation with NOAA and the U.S. Coast Guard (USCG), shall provide assistance to countries in priority regions and priority flag states to improve effectiveness of IUU fishing enforcement, including through law enforcement trainings and coordination activities.[144]
- The Department of State, in consultation with NOAA and the USCG, shall support countries in priority regions and priority flag states in adopting and implementing the PSMA.[145]
- The Department of State, in consultation with NOAA and the USCG, shall help countries in priority regions and priority flag states increase their capacity for IUU fishing investigations and prosecutions.[146]

---

[140] 16 U.S.C. §8002.
[141] 16 U.S.C. §8031. For more information, see the "What Is the Interagency Working Group on IUU Fishing?" section of this report.
[142] 16 U.S.C. §8011.
[143] 16 U.S.C. §8012.
[144] 16 U.S.C. §8013(b). For more information, see the "What Is the Interagency Working Group on IUU Fishing?" section of this report.
[145] 16 U.S.C. §8013(c).
[146] 16 U.S.C. §8013(d). For more information, see the "What Is the Interagency Working Group on IUU Fishing?" section of this report.

- Relevant agencies (e.g., the Department of Defense [DOD], NOAA, the USCG) shall expand mechanisms to combat IUU fishing, such as entering into shiprider agreements.[147]
- Relevant agencies (e.g., the Department of State, NOAA, the U.S. Agency for International Development [USAID]) shall work to improve transparency and traceability programs, including sharing knowledge with countries in priority regions and priority flag states.[148]
- Relevant agencies (e.g., the Department of State, NOAA, USAID, the USCG) shall expand the role of technology in combatting IUU fishing.[149]

Other U.S. laws may address aspects of IUU fishing. For example, the Pelly Amendment to the Fishermen's Protective Act (P.L. 92-219) provides the President with the authority to limit the importation of any products from a nation where its nationals are engaging in trade or other activities that diminish the effectiveness of any international conservation program for threatened or endangered species or international fisheries.[150] Congress may consider whether U.S. laws (e.g., the Pelly Amendment, the Moratorium Protection Act) provide an adequate means to identify and sanction vessels, companies, or countries that participate or condone IUU activities or if such efforts could be strengthened.[151] Congress also may consider whether certain U.S. laws should be expanded to include other types of illegal activities associated with fishing operations, such as human trafficking, or whether these concerns should be handled through other existing laws, such as the Tariff Act of 1930 or the Trafficking Victims Protection Act.[152]

---

[147] 16 U.S.C. §8014(a). For more information on shiprider agreements, see the "What Are Shipriders?" section of this report.

[148] 16 U.S.C. §8015.

[149] 16 U.S.C. §8016. For more information, see the "What Technologies Can Be Used to Identify Vessels Suspected of IUU Fishing?" section of this report.

[150] 22 U.S.C. §1978.

[151] For example, NOAA, NMFS, "NOAA Engagement with Nations and Entities Under the Moratorium Protection Act"; and NOAA, NMFS, "Port Restrictions Under the Moratorium Protection Act," https://www.fisheries.noaa.gov/ content/port-restrictions-under-moratorium-protection-act (hereinafter NOAA, NMFS, "Port Restrictions Under the Moratorium Protection Act").

[152] 19 U.S.C. §§1202-1683g; and 22 U.S.C. §§7101-7115.

What Is the Interagency Working Group on IUU Fishing?

The Maritime SAFE Act established the Interagency Working Group (IWG) on IUU Fishing to support and coordinate a government-wide effort to address IUU fishing globally.[153] The working group comprises representatives from 21 federal agencies.[154] In June 2023, the U.S. Department of State started its three-year term as chair of IWG on IUU Fishing, with representatives from NOAA and the USCG serving as deputy chairs.[155] The chair of the working group rotates among the Secretary of the Department of Homeland Security (in which the USCG operates), Secretary of State, and NOAA Administrator.[156]

Congress directed the IWG on IUU Fishing to develop a "strategic plan for combating IUU fishing and enhancing maritime security, including specific strategies with monitoring benchmarks for addressing IUU fishing in priority regions."[157] In October 2022, the working group released its National 5-Year Strategy for Combatting IUU Fishing.[158] The strategy includes three objectives to combat IUU fishing: (1) promote sustainable fisheries management and governance; (2) enhance the monitoring, control, and surveillance of marine fishing operations; and (3) ensure only legal, sustainable, and responsibility harvested seafood enters trade.[159]

Congress also charged the IWG on IUU Fishing to identify priority regions and priority flag states. A priority region means a region "(A) that is at high risk for IUU fishing activity or the entry of illegally caught seafood into the markets of countries in the region; and (B) in which countries lack the capacity to fully address the illegal activity described in subparagraph (A)."[160]

---

[153] Ibid.; and NOAA, NMFS, "U.S. Interagency Working Group on IUU Fishing," https://www.fisheries.noaa.gov/national/international-affairs/us-interagency-working-group-iuu-fishing. Hereinafter, NOAA, NMFS, "U.S. Interagency Working Group on IUU Fishing."

[154] 16 U.S.C. §8031(b). For a list of the 21 agencies, see NOAA, NMFS, "U.S. Interagency Working Group on IUU Fishing."

[155] U.S. Department of State, "Illegal, Unreported, and Unregulated Fishing." NOAA served as the first chair for the U.S. Interagency Working Group (IWG) on IUU Fishing (NOAA, NMFS, "U.S. Interagency Working Group on IUU Fishing").

[156] 16 U.S.C. §8031(b)(1).

[157] 16 U.S.C. §8032.

[158] IWG on IUU Fishing, *National 5-Year Strategy for Combating Illegal, Unreported, and Unregulated Fishing: 2022- 2026*, Report to Congress, October 2022, pp. 1-A3-1. Hereinafter IWG on IUU Fishing, *National 5-Year Strategy for Combating IUU Fishing*.

[159] Ibid., p. 6.

[160] 16 U.S.C. §8001(9).

The IWG on IUU Fishing assessed different regions and placed regions into three tiers of priority (Table 2).

- Tier One represents "regions where there was both clear information about the challenges resulting from IUU fishing and ample existing opportunities for U.S. partnerships and activities that could address those challenges."[161]
- Tier Two represents regions where "U.S. agencies and our partners are looking for opportunities to build law enforcement cooperation, share information, and support training and capacity building within these regions."[162]
- Tier Three represents regions where IUU fishing has been raised as a concern, though "details are limited."[163]

Table 2. Priority regions at risk for IUU fishing

(as identified by the U.S. Interagency Working Group on IUU Fishing)

Tier One	Tier Two	Tier Three
South and Central America (Pacific Ocean)	Central America and Caribbean (Gulf of Mexico, Caribbean Sea)	Middle East and Gulf States (Persian Gulf, Gulf of Oman, Gulf of Aden, Red Sea)
Gulf of Guinea	South America (Atlantic Ocean)	South Asia (Bay of Bengal)
South Asia (Gulf of Thailand, Java Sea, Banda Sea, Celebes Sea)	Northwest Africa (Atlantic Ocean)	East Asia Pacific (East China Sea, Sea of Japan, Sea of Okhotsk)
Pacific Islands	Southern and Central Africa (Atlantic and Indian Ocean)	—
—	East Africa (Indian Ocean)	—

Source: U.S. Interagency Working Group on IUU Fishing, National 5-Year Strategy for Combating Illegal, Unreported, and Unregulated Fishing: 2022-2026, Report to Congress, October 2022, pp. A1-1–A1-3.
Notes: IUU fishing = illegal, unreported, and unregulated fishing.

The IWG on IUU Fishing identified *priority flag* states or authorities based on those with vessels that "actively engage in, knowingly profit from, or are complicit in IUU fishing" and the priority flag states or authorities are "willing, but lack ... the capacity, to monitor or take effective enforcement action against ... [their fleets]."[164] Based on this definition, the IWG on IUU

---

[161] IWG on IUU Fishing, *National 5-Year Strategy for Combating IUU Fishing*, p. A1-2.
[162] Ibid.
[163] Ibid., p. A1-3.
[164] 16 U.S.C. §8001(8).

Fishing identified five priority flag states and authorities to work with: Ecuador, Panama, Senegal, Taiwan, and Vietnam (Figure 3).[165]

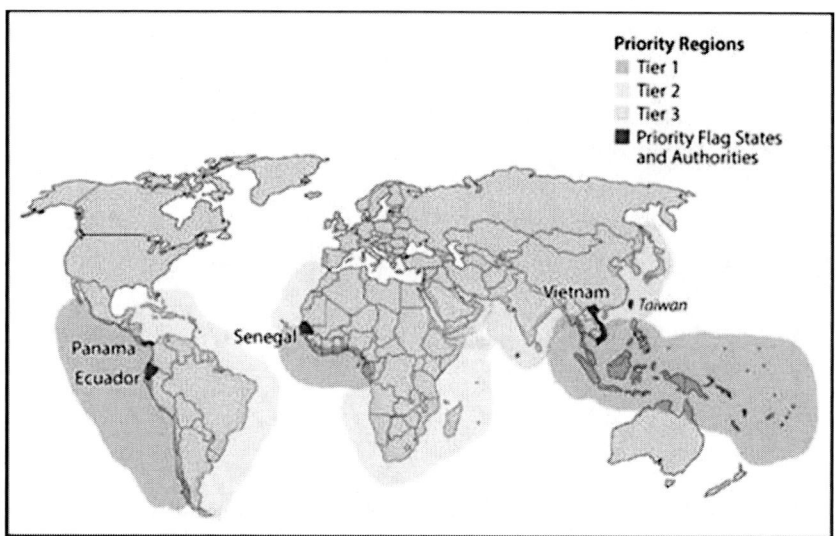

Source: Congressional Research Service, modified from U.S. Interagency Working Group on IUU Fishing, National 5-Year Strategy for Combating Illegal, Unreported, and Unregulated Fishing: 2022-2026, Report to Congress, October 2022, p. A2-1.

Notes: IUU fishing = illegal, unreported, and unregulated fishing.

Figure 3. Priority Flag States and Authorities Overlaid with Priority Regions. (as identified by the U.S. Interagency Working Group on IUU Fishing).

Congress also directed the IWG on IUU Fishing to submit a report to Congress no later than five years after the IWG's submission of the National 5-Year Strategy for Combatting IUU Fishing. The report is required to contain summaries of global and regional trends in IUU fishing and situational threats with respect to IUU fishing in priority regions and the capacity of countries in the regions to respond to such threats as a result of U.S. assistance, among other summaries.[166] The report also is required to assess the extent of the convergence of transnational crimes (i.e., human trafficking and forced labor) and IUU fishing; the capacity of priority flag states to police their fleet; and

---

[165] IWG on IUU Fishing, *National 5-Year Strategy for Combating IUU Fishing*, p. A2-1.
[166] 16 U.S.C. §8033.

the involvement of organizations designated as foreign terrorist organizations in IUU fishing, among other assessments outlined in the national strategy.[167]

Congress also directed NOAA, in coordination with the Department of State and the USCG, to establish an IWG sub-working group to address IUU fishing in the U.S. EEZ in the Gulf of Mexico.[168] Pursuant to the Maritime SAFE Act, NOAA submitted a report to Congress in 2021 about federal actions and policies to address "Mexican nationals operating out of fishing camps in Tamaulipas state, repeatedly entering the U.S. [EEZ] of the Gulf of Mexico via small boats, and fishing without authorization."[169] These findings were in addition to information included in NOAA's recent reports to Congress regarding Mexico IUU fishing activities submitted in accordance with the Moratorium Protection Act.[170]

What Actions Are U.S. Agencies Taking to Address IUU Fishing?

Several federal departments and agencies, including DOD, the Department of State, NOAA, and the USCG, engage in various efforts to combat IUU fishing on the high seas and in the EEZs of partner nations. These agencies' efforts include establishing partnerships; improving enforcement tools, such as HSBI; identifying and sharing information about countries that have fishing vessels participating in IUU fishing activities; participating in joint investigations of IUU fishing activities; and assisting partner nations to develop and maintain their own counter IUU fishing capacity, among other lines of effort.[171]

---

[167] Ibid.
[168] 16 U.S.C. §8034.
[169] NOAA, *Report of the Gulf of Mexico Illegal, Unreported, and Unregulated Fishing Subworking Group*, Report to Congress, 2021.
[170] NOAA, NMFS, Improving International Fisheries Management: 2019 Report to Congress, September 2019; NOAA, NMFS, Improving International Fisheries Management: 2021 Report to Congress, August 2021; NOAA, NMFS, "NOAA Engagement with Nations and Entities Under the Moratorium Protection Act"; and NOAA, NMFS, "Port Restrictions Under the Moratorium Protection Act." For more information, see the "Magnuson-Stevens Fishery Conservation and Management Reauthorization Act of 2006 and High Seas Driftnet Fishing Moratorium Protection Act" section of this report.
[171] NOAA, NMFS, *Improving International Fisheries Management*, Report to Congress, August 2023, p. 3.

Selected efforts from federal agencies (in alphabetical order) are described below.[172]

National Oceanic and Atmospheric Administration. NOAA NMFS coordinates with federal agencies, foreign governments, international organizations, and other partners to address IUU fishing. NMFS identifies nations and entities that have vessels participating in IUU fishing activities and fishing activities that result in bycatch (i.e., nontarget catch) of protected species or sharks on the high seas or in any nation's EEZ.[173] NMFS also conducts PSMA inspections and enforces multilateral agreements and regulations.[174] Furthermore, NMFS trains personnel from other nations to enhance their abilities to implement the PSMA and participates in joint capacity- building workshops through international partnerships to address IUU fishing.[175] NMFS and U.S. Customs and Border Protection (CBP) monitor U.S. seafood imports through the U.S. Seafood Import Monitoring Program.[176] Additionally, NMFS participates in other trade monitoring programs, such as the Antarctic Marine Living Resources Program, Atlantic Highly Migratory Species International Trade Program, and Tuna Tracking and Verification Program. NMFS supports Fisheries International Cooperation Projects, including those focused on addressing IUU fishing, through the Fisheries International Cooperation and Assistance Program. Additionally, NMFS enforces other related laws, such as the Lacey Act, as amended,[177] which prohibits the sale or purchase of any wildlife taken or sold in violation of any U.S. law, treaty, or regulation, and prohibits false labeling.[178]

---

[172] For a more comprehensive list of federal departments and agencies involved in addressing IUU fishing, see IWG on IUU Fishing, "Working Group Member Agencies," https://iuufishing.noaa.gov/member-agencies/.

[173] NOAA, NMFS, "NOAA Engagement with Nations and Entities Under the Moratorium Protection Act"; see footnote 133.

[174] In accordance with 16 U.S.C. §8013(c).

[175] In accordance with 16 U.S.C. §8013(b); NOAA, NMFS, "Countering Illegal, Unreported, and Unregulated Fishing: Capacity Building and Technical Assistance," https://www.fisheries.noaa.gov/enforcement/countering-illegal-unreported-and-unregulated-fishing-capacity-building-and-technical#counter-iuu-fishing-technical-assistance-and- capacity-building.

[176] For more information, see the "What Is the Seafood Import Monitoring Program?" section of this report.

[177] 16 U.S.C. §§3371-3378 and 18 U.S.C. §§42-43.

[178] NOAA, NMFS, "Understanding Laws and NOAA Fisheries – What is the Lacey Act and why is it important," https://www.fisheries.noaa.gov/insight/understanding-laws-and-noaa-fisheries#what-is-the-lacey-act-and-why-is-it-important?; U.S. Department of Justice, Environment and Natural Resources Division, "Environmental Crimes Bulletin – July 2024," https://www.justice.gov/enrd/blog/environmental-crimes-bulletin-july-2024; U.S.

U.S. Agency for International Development. USAID administers biodiversity programs to promote marine conservation and sustainable fisheries management, which include activities to combat IUU fishing.[179] Some of USAID's efforts to combat IUU fishing focus on promoting seafood traceability.[180] For example, USAID works with several private foundations through the Seafood Alliance for Legality and Traceability to help governments and communities to promote seafood traceability and adopt a digital traceability system.[181] In addition, USAID's Feed the Future initiative may approach IUU fishing from a food security perspective. For example, a project in Senegal funded under USAID's Feed the Future initiative seeks to curb overfishing and the use of illegal fishing equipment and practices, among other aims.[182] In addition to these efforts, S.Rept. 118-200, the Appropriations Committee report accompanying S. 4797 in the 118th Congress, would direct USAID's Bureau for Inclusive Growth, Partnership, and Innovation and Bureau for Resilience, Environment, and Food Security to "work together to address the fundamental system failures that allow for IUU fishing to persist, jeopardizing economic, environmental, and food security objectives."

U.S. Coast Guard. The USCG is a multi-mission maritime service with the authority to conduct maritime law enforcement operations, including operations aimed at combating IUU fishing activity.[183] The USCG enforces U.S. and international living marine resources laws in the U.S. EEZ and in key areas of the high seas. The USCG counters IUU fishing using measures such

---

Fish and Wildlife Service (FWS), "Lacey Act Amendments of 1981," https://www.fws.gov/law/lacey-act-amendments-1981.

[179] IWG on IUU Fishing, "United States Agency for International Development," https://iuufishing.noaa.gov/member-agencies/usagencyforinternationaldevelopment/.
According to the U.S. Agency for International Development (USAID), in FY2021, the agency "invested more than $60 million in over 20 countries to promote sustainable fisheries and conserve marine biodiversity" (USAID, "Illegal, Unreported, and Unregulated Fishing," https://www.usaid.gov/ biodiversity/illegal-unreported-and-unregulated-fishing).

[180] USAID, "Illegal, Unreported, and Unregulated Fishing," https://www.usaid.gov/biodiversity/illegal-unreported-and- unregulated-fishing.

[181] FishWise, "SALT," https://fishwise.org/salt/.

[182] Feed the Future, "Senegal Dekkal Geej, Towards Sustainable Fisheries," https://winrock.org/wp-content/uploads/2019/06/20200103-FtF-Senegal-Dekkal-Geej-Handout.pdf; and IWG on IUU Fishing, *National 5-Year Strategy for Combating IUU Fishing*, p. 11.

[183] 14 U.S.C. §102.

as at-sea operations, vessel tracking data,[184] shiprider agreements,[185] and cooperation in partner nation capacity-building exercises.[186]

The USCG has limited legislative authority to unilaterally provide training and technical assistance to foreign countries to address IUU fishing.[187] The USCG is generally considered a service provider under auspices of security assistance and cooperation programs under Title 22 and Title 10 of the *U.S. Code*. Title 10, Section 301(7) of the *U.S. Code* defines *security cooperation programs and activities of the Department of Defense* as any program, activity (including exercise), or interaction of the Department of Defense with the security establishment of a foreign country to achieve a purpose as follows:

a) To build and develop allied and friendly security capabilities for self-defense and multinational operations.
b) To provide the armed forces with access to the foreign country during peacetime or a contingency operation.
c) To build relationships that promote specific United States security interests.

The USCG may, when requested to do so by another federal agency (e.g., Department of State, DOD), use its personnel and facilities to participate in foreign security assistance and cooperation activities.[188] The USCG must be "especially qualified" to provide such requested capacity-building assistance, according to the law.[189]

The USCG also could provide technical assistance, including law enforcement and maritime safety and security training, to foreign navies, coast guards, and other maritime law enforcement agencies, including national-level security forces.[190] Under this scenario, another federal agency's international engagement authorities would be conveyed to the USCG with the transfer of funding.

---

[184] For more information, see the "What Technologies Can Be Used to Identify Vessels Suspected of IUU Fishing?" section of this report.
[185] For more information, see the "What Are Shipriders?" section of this report.
[186] IWG on IUU Fishing, *National 5-Year Strategy for Combating IUU Fishing*, pp. 1-A3-1; and U.S. Coast Guard (USCG), *Illegal, Unreported, and Unregulated Fishing Strategic Outlook Implementation Plan*, July 2021, pp. 1-29.
[187] Email correspondence from the USCG to CRS, June 14, 2024.
[188] 14 U.S.C. §701.
[189] 14 U.S.C. §701(a).
[190] 14 U.S.C. §710(b).

U.S. Department of Defense. DOD supports federal agencies and foreign partners involved in directly combatting IUU fishing.[191] Congress has authorized DOD to use its appropriations to fund and conduct security cooperation activities with national-level security forces, generally under Title 10 of the *U.S. Code.*[192] Most DOD security cooperation funding is authorized under DOD's "building partner capacity" authority, which authorizes DOD to bolster maritime security capacities, among other activities.[193] DOD also implements some Department of State security assistance through the provision of training and defense equipment for partner navies and national-level security forces via programs such as Excess Defense Article and Foreign Military Sales. The House-passed version of the Servicemember Quality of Life Improvement and National Defense Authorization Act for Fiscal Year 2025 (H.R. 8070) would amend 10 U.S.C. §333(a) to include counter-illegal, unreported, and unregulated fishing operations as part of the Secretary of Defense's foreign capacity building authorities.[194] Also, H.Rept. 118-529, the House Armed Services Committee report accompanying its reported version of H.R. 8070, would direct DOD to provide the committee an "overview of fishing activities in the Indo-Pacific region by the distant-water fishing fleets of foreign governments that are employed as extensions of such countries' official maritime security forces."[195]

U.S. Department of State. According to the Department of State, the department is working to (1) strengthen overall ocean governance and make multilateral processes more effective; (2) increase fishing transparency requirements, improve information sharing across the U.S. government and with allied and partner nations, and implement cooperative enforcement and penalty tools; (3) apply innovative technologies to identify IUU fishing; and (4) raise awareness and commitments of collaborators to counter IUU fishing.[196] The Department of State also administers security assistance programs in foreign countries that (among other objectives) aim to build the capacity of maritime law enforcement agencies to govern their maritime domains, including fishery enforcement.[197] Both the Department of State and some Members of Congress have recognized China's role in the exploitation

---

[191] IWG on IUU Fishing, "Department of Defense," https://iuufishing.noaa.gov/member-agencies/departmentofdefense/.
[192] For example, 10 U.S.C. §§311, 312, 321, 333.
[193] 10 U.S.C. §333.
[194] See Section 1237 of H.R. 8070 in the 118th Cong.
[195] See Title XIII of H.Rept. 118-529 in the 118th Cong.
[196] U.S. Department of State, "Illegal, Unreported, and Unregulated Fishing."
[197] For example, 22 U.S.C. §§2348, 2291, 2763.

of global fisheries,[198] including through requesting and recommending funding to the Department of State's "Countering [People's Republic of China] Influence Fund" to address IUU fishing threats.

U.S. Fish and Wildlife Service. The U.S. Fish and Wildlife Service's (FWS's) Office of Law Enforcement attaché program sends special agents to U.S. embassies in host nations to coordinate with host country officials in wildlife trafficking and natural resource criminal investigations.[199] The 2013 Executive Order 13648, "Combatting Wildlife Trafficking," initiated the FWS attaché program, among other actions to address wildlife and natural resource crimes.[200] Historically, FWS attachés have worked on terrestrial (e.g., elephant, rhino) wildlife trafficking crimes. In recent years, IUU fishing issues have become more pressing,[201] leading FWS to increase its engagement on IUU fishing issues. For example, FWS has conducted seminars with host nations to increase their awareness of IUU fishing and aid in criminal investigations. In 2019, the FWS attaché established a Gabonese IUU fishing pilot project to detect IUU fishing activities.[202] The project used open-source information to identify vessels and their movement patterns in the Gabonese EEZ and the broader West African Coast. One outcome of the project was the interdiction of several vessels, including the seizure of a Chinese fishing trawler that was a repeated illegal fishing offender in Gabonese water. FWS also enforces the Lacey Act, as amended, a mechanism for implementing trade restrictions on the import and illegal trade of certain wildlife (including fish), plants, and related products, and the Convention on International Trade in Endangered Species of Wild Fauna and Flora.[203]

---

[198] For example, see S.Rept. 118-200, the Senate Committee on Appropriations report accompanying S. 4797, and H.Rept. 118-554, the House Committee on Appropriations report accompanying H.R. 8771, in the 118th Cong.; and U.S. Department of State, *Congressional Budget Justification: Department of State, Foreign Operations, and Related Programs Fiscal Year 2025*, p. 125.

[199] IWG on IUU Fishing, "U.S. Fish and Wildlife Service," https://iuufishing.noaa.gov/member-agencies/ usfishandwildlifeservice/; FWS, "International Affairs—Our Laws and Regulations," https://www.fws.gov/program/ international-affairs; and Executive Order 13773, "Enforcing Federal Law with Respect to Transnational Criminal Organizations and Preventing International Trafficking," 82 *Federal Register* 10691, February 14, 2017.

[200] Executive Order 13648, "Combatting Wildlife Trafficking," 78 *Federal Register* 40621, July 5, 2013; and White House, *National Strategy for Combatting Wildlife Trafficking*, February 2014, pp. 1-12. The Eliminate, Neutralize, and Disrupt Wildlife Trafficking Act of 2016 (P.L. 114-231) established a national policy on wildlife trafficking. 16 U.S.C.§7643.

[201] Email correspondence from FWS to CRS, November 1, 2023.

[202] U.S. Department of State, "2020 END Wildlife Trafficking Strategic Review," October 26, 2020, https://www.state.gov/2020-end-wildlife-trafficking-strategic-review/.

[203] As of September 2024, the Convention on International Trade in Endangered Species of Wild Fauna and Flora (CITES) regulates the trade of more than 34,300 species of plants and 6,600

What Is the Seafood Import Monitoring Program?

NMFS and CBP monitor U.S. seafood imports through the Seafood Import Monitoring Program (SIMP) with the goal of preventing imported IUU fish and fish products from entering U.S. commerce.[204] SIMP is a risk-based seafood traceability program that sets reporting and recordkeeping requirements from the point of harvest to entry into U.S. commerce.[205] The program requires seafood importers to exercise increased control over their supply chains, particularly for 13 seafood species groups,[206] by obtaining an NMFS International Fisheries Trade Permit.[207] SIMP also requires U.S. importers to retain chain-of-custody information for all imports of covered species and to document each step of the supply chain.[208] The program currently covers nearly half of all U.S. seafood imports, comprising approximately 1.7 billion pounds of seafood as of FY2023,[209] with data for priority species collected through the International Trade Data System.[210] SIMP does not require importers to collect labor information.[211]

SIMP was established by the Secretary of Commerce in 2016, and its implementation began in 2018 in accordance with the Consolidated Appropriations Act, 2018 (P.L. 115-141). The program aims to serve as both a screening mechanism and a deterrent against the entry of IUU-associated and misrepresented seafood into the U.S. market.[212] SIMP does not include

---

species of animals (CITES, "The CITES Species," https://cites.org/eng/disc/species.php). The United States is a party to CITES (CITES, "List of Parties to the Convention," https://cites.org/eng/disc/parties/index.php).

[204] NOAA, NMFS, "Seafood Import Monitoring Program," https://www.fisheries.noaa.gov/international/international-affairs/seafood-import-monitoring-program. Hereinafter NOAA, NMFS, "Seafood Import Monitoring Program."

[205] NOAA, NMFS, "Seafood Import Monitoring Program Facts and Reports," https://www.fisheries.noaa.gov/international/international-affairs/seafood-import-monitoring-program-facts-and-reports. Hereinafter NOAA, NMFS, "Seafood Import Monitoring Program Facts and Reports."

[206] NOAA, NMFS, "Seafood Import Monitoring Program."

[207] NOAA, NMFS, "International Fisheries Trade Permit," https://www.fisheries.noaa.gov/permit/international-fisheries-trade-permit.

[208] NOAA, NMFS, "Seafood Import Monitoring Program Facts and Reports."

[209] Ibid.; NOAA, NMFS, *Report on the Seafood Import Monitoring Program—FY2023.*

[210] Examples of data include information on species, quantity, vessel and flag state, fishing gear used, landing or offloading dates, names of recipients, and U.S. importer of record. NOAA, NMFS, *Report on the Seafood Import Monitoring Program—FY2023.*

[211] NOAA, NMFS, *Report on the Implementation of the U.S. Seafood Import Monitoring Program*, April 2021, pp. 1-17. Hereinafter NOAA, NMFS, *Report on the Implementation of the U.S. Seafood Import Monitoring Program.*

[212] Ibid.

routine examinations of each shipment but relies on data reporting and recordkeeping requirements as well as random and targeted audits of shipments performed by NMFS, CBP, and state agency partners.[213]

Congress requires NMFS to submit reports regarding its efforts through SIMP to prevent seafood associated with IUU fishing and forced labor from entering the United States, including annual reports on the program.[214] Each annual report is to include information regarding the volume and value of seafood species subject to SIMP, NMFS enforcement efforts related to SIMP, the percentage of import shipments audited, instances of noncompliance with SIMP requirements, and seafood species and harvest locations for which noncompliance and legal violations were most prevalent, among other information.[215]

NMFS reported in 2024 that it had undertaken over 3,900 audits of seafood imports from January 2018 through September 2023, which comprised 0.5% of all SIMP imports since implementation.[216] The majority of SIMP imports typically include shrimp (33%-56%), tunas (31%-47%), and cods (5%-10%), based on reported FY2020 and FY2023 weight values.[217] Proportions of total audits by species group generally align with these percentages.[218] NMFS identified noncompliance with SIMP requirements in approximately 43% of audited scafood during FY2020 and noncompliance in 56% of audits during FY2023.[219] Compliance with SIMP requirements per species may vary, and incidents of noncompliance may be greater in less frequently audited species (e.g., 16 audits of blue crab were conducted in FY2020, of which 88% were noncompliant, compared with 448 audits of shrimp that year, of which 35% were noncompliant). During FY2023, incomplete chain of custody and misreported harvest weight were the most frequent findings in noncompliant audits, with tunas having the highest

---

[213] Ibid.; NOAA, NMFS, *Report on the Seafood Import Monitoring Program—FY2023*.
[214] For example, as stipulated in the Consolidated Appropriations Act, 2018 (P.L. 115-141); Consolidated Appropriations Act, 2020 (P.L. 116-93); and Don Young Coast Guard Authorization Act of 2022 (Division K, Title CXIII of P.L. 117-263); 16 U.S.C. §1885a.
[215] 16 U.S.C. §1885a(b).
[216] NOAA, NMFS, *Report on the Seafood Import Monitoring Program—FY2023*.
[217] Ibid.; NOAA, NMFS, Report on the Implementation of the U.S. Seafood Import Monitoring Program.
[218] NOAA, NMFS, Report on the Implementation of the U.S. Seafood Import Monitoring Program; and NOAA, NMFS, Report on the Seafood Import Monitoring Program—FY2023.
[219] NOAA, NMFS, Report on the Implementation of the U.S. Seafood Import Monitoring Program; and NOAA, NMFS, Report on the Seafood Import Monitoring Program—FY2023.

frequency of noncompliance. In cases of noncompliance, only a small number typically warrant enforcement action by NMFS's Office of Law Enforcement (e.g., approximately 10% of SIMP- related cases initiated in FY2020 resulted in civil penalties).[220]

NMFS issued a proposed rule in December 2022 to expand SIMP in accordance with directives in the June 2022 White House Memorandum on Combatting IUU Fishing and Associated Labor Abuses.[221] NMFS proposed adding species or groups of species beyond the 13 currently associated with SIMP.[222] In November 2023, NMFS announced it was withdrawing this proposed rule in light of the public comments it received and would instead conduct a comprehensive program review "to strengthen the impact and effectiveness of SIMP."[223] NMFS completed this program review in summer 2024 and plans to implement its recommendations.[224] The June 2022 White House memorandum also directed NMFS to seek resources and technologies to improve SIMP's effectiveness. NMFS has stated that it intends to update its current audit procedures to incorporate automated screening, including models

---

[220] NOAA, NMFS, Report on the Implementation of the U.S. Seafood Import Monitoring Program; and NOAA, NMFS, Report on the Seafood Import Monitoring Program— FY2023.

[221] NOAA, NMFS, "Magnuson-Stevens Fishery Conservation and Management Act; Seafood Import Monitoring Program," 87 Federal Register 79836-79848, December 28, 2022 (hereinafter NOAA, NMFS, 87 Federal Register 79836-79848); and White House, Memorandum on Combating Illegal, Unreported, and Unregulated Fishing and Associated Labor Abuses, National Security Memorandum/NSM-11, June 27, 2022.

[222] These species and species groups would have included all species of snappers (family Lutjanidae); several additional tuna species; and new species groups of cuttlefish, squid, octopus, eels (Anguilla sp.), queen conch (Aliger gigas), and Caribbean spiny lobster (Panulirus argus). NOAA, NMFS, 87 Federal Register 79836-79848.

[223] NOAA, NMFS, "Magnuson-Stevens Fishery Conservation and Management Act; Seafood Import Monitoring Program," 88 Federal Register 78714-78715, November 16, 2023; and NOAA, NMFS, "NOAA Fisheries Announces Comprehensive Review of Its Seafood Import Monitoring Program to Strengthen Its Impact and Effectiveness," https://www.fisheries.noaa.gov/feature-story/noaa-fisheries-announces-comprehensive-review-its-seafood-import- monitoring-program.

[224] NOAA, NMFS, "Webinars on the Seafood Import Monitoring Program: Comprehensive Review Update," https://www.fisheries.noaa.gov/event/webinars-seafood-import-monitoring-program-comprehensive-review-update; Sally Yozell et al., *Workshop Summary Report: Reimagining the Seafood Import Monitoring Program, Workshop I*, Stimson Center and FishWise, May 2024, pp. 1-32, https://www.stimson.org/wp-content/uploads/2024/07/ REPORT_ReimaginingSIMPWorkshop1StimsonFishWise.pdf; and Sally Yozell et al., *Workshop Summary Report: Reimagining the Seafood Import Monitoring Program, Session II*, Stimson Center and FishWise, July 2024, pp. 1-27, https://www.stimson.org/wp-content/uploads/2024/07/REPORT_Reimagining-SIMP-Session-2-StimsonFishWise.pdf.

that incorporate artificial intelligence and machine learning approaches, as also directed by Congress.[225]

Some U.S. programs, such as SIMP and existing customs enforcement measures, have attempted to provide greater scrutiny of seafood imports. To support these efforts, Congress has provided increasing funding for SIMP since its implementation, most recently providing approximately $6 million per year in FY2024.[226] Some experts and stakeholders have characterized NMFS's implementation of SIMP as "a good start," while recommending additional international coordination and broader expansions of the program beyond those initially proposed by NMFS.[227] Other stakeholders have raised concerns that SIMP fails to address human rights violations in the seafood industry and includes gaps in traceability from the point of import to the final point of sale.[228] Several stakeholders have questioned the overall effectiveness of SIMP, because some Americans are still consuming seafood associated with IUU fishing.[229] Stakeholders and experts also point out the need for greater enforcement capacity for traceability programs such as SIMP to be effective in preventing IUU fishing products from entering national markets.[230]

Congress may consider expanding SIMP to include all species imported by the United States. Congress also may consider increasing customs

---

[225] NOAA, NMFS, *Report on the Seafood Import Monitoring Program—FY2023*; "Explanatory Statement Submitted by Ms. DeLauro, Chair of the House Committee on Appropriations, Regarding the House Amendment to the Senate Amendment to H.R. 2471, Consolidated Appropriations Act, 2022," *Congressional Record*, vol. 168, No. 42-Book III (March 9, 2022), p. H1778.

[226] "Explanatory Statement Submitted by Mrs. Murray, Chair of the Senate Committee on Appropriations, Regarding the H.R. 4366, Consolidated Appropriations Act, 2024," *Congressional Record*, vol. 170, No. 39 (March 5, 2024), p. S1401.

[227] Jessica A. Gephart, Halley E. Froehlich, and Trevor A. Branch, "Opinion: To Create Sustainable Seafood Industries, the United States Needs a Better Accounting of Imports and Exports," *Proceedings of the National Academy of Sciences*, vol. 116, no. 19 (2019), pp. 9142-9146; Natural Resources Defense Council, *Strengthening U.S. Leadership to Deter Illegal Seafood: Implementation Challenges and Recommendations for the Seafood Import Monitoring Program*, January 2023.

[228] Jack Cheney, "What Is the Seafood Import Monitoring Program (SIMP)?," University of Washington, April 12, 2022, https://sustainablefisheries-uw.org/simp-seafood-import-monitoring-program/.

[229] Ibid.; National Fisheries Institute, "The Modern-Day Lawn Dart: NOAA's Seafood Import Monitoring Program," https://aboutseafood.com/the-modern-day-lawn-dart-noaas-seafood-import-monitoring-program/.

[230] Catherine S. Longo et al., "A Perspective on the Role of Eco-Certification in Eliminating Illegal, Unreported, and Unregulated Fishing," *Frontiers in Ecology and Evolution*, vol. 9 (2021), 637228, pp. 1-14; and Donovan, "Role of Corporations."

enforcement at the border and whether greater resources are needed to fully account for IUU fished seafood entering the United States.

Increased efforts to trace seafood also serve multiple purposes, such as improving seafood safety, stopping seafood fraud, and identifying seafood production related to human trafficking. In addition, Congress may explore ways to improve seafood traceability to verify information currently collected on the origin of and route taken by seafood before its entry into the United States.

What Are Shipriders?

Some maritime law enforcement agreements include shiprider provisions that authorize a law enforcement official of one party to embark on a law enforcement vessel or aircraft of the other party and exercise certain authorities. An agreement that includes a shiprider provision is commonly referred to as a shiprider agreement. U.S. shiprider agreements are designed to allow U.S. law enforcement officials to assist partner nations in combatting various illicit maritime activity, such as IUU fishing and trafficking in narcotic drugs and psychotropic substances. In general, U.S. bilateral shiprider agreements allow maritime law enforcement officers of the partner nation to embark on warships and other vessels (and/or aircraft) of the U.S. government. The presence of a shiprider on board a U.S. government vessel allows the vessel to enforce the laws and regulations of the partner nation, including the observation and investigation (i.e., board and search) of suspect vessels, within the partner nation's designated territorial sea or EEZ. Certain shiprider agreements also allow U.S. government vessels with embarked shipriders to pursue flag ships of the party on the high seas.

Not all U.S. shiprider agreements include counter-IUU fishing provisions. One priority of the IWG on IUU Fishing is for the U.S. government to establish new bilateral shiprider agreements that have counter-IUU fishing provisions with countries located within priority regions and to add counter-IUU fishing provisions to existing shiprider agreements.[231] The United States has entered into bilateral shiprider agreements to address IUU fishing with

---

[231] The Maritime SAFE Act directs selected federal officials to "exercise existing shiprider agreements and to enter into and implement new shiprider agreements" (16 U.S.C. §8013(b)(2)).

several nations, including Cook Islands,[232] Côte d'Ivoire,[233] Ecuador,[234] Fiji,[235] Gambia,[236] Kiribati,[237] Micronesia,[238] Nauru,[239] Palau,[240] Panama,[241]

---

[232] Agreement Between the Government of the United States of America and the Government of the Cook Islands Concerning Cooperation in Joint Maritime Surveillance Operations (T.I.A.S. 08-725), signed July 25, 2008.

[233] Agreement Between the United States of America and the Republic of Cote D'Ivoire Concerning Counter Illicit Transnational Maritime Activity Operations (T.I.A.S. 24-206), signed February 6, 2024.

[234] Agreement Between the United States of America and the Republic of Ecuador Concerning Counter Illicit Transnational Maritime Activity Operations (T.I.A.S. 24-223), signed September 27, 2023.

[235] Agreement Between the Government of the Republic of Fiji and the Government of the United States of America Concerning Counter Illicit Transnational Maritime Activity Operations (T.I.A.S. 18-1112), signed November 12, 2018.

[236] Agreement Between the Government of the United States of America and the Government of the Republic of the Gambia Concerning Cooperation to Suppress Illicit Transnational Maritime Activity (T.I.A.S. 11-1010), signed October 10, 2011.

[237] Agreement Between the Government of the United States of America and the Government of the Republic of Kiribati Concerning Cooperation in Joint Maritime Surveillance Operations (T.I.A.S. 08-1124), signed November 24, 2008.

[238] Agreement Between the Government of the United States of America and the Government of the Federated States of Micronesia Concerning Cooperative Shiprider Agreement (T.I.A.S. 08-514), signed May 14, 2008; and Agreement Between the Government of the United States of America and the Government of the Federated States of Micronesia Concerning Operational Cooperation to Suppress Illicit Transnational Maritime Activity (T.I.A.S. 14-303), signed March 3, 2014.

[239] Agreement Between the Government of the United States of America and the Government of the Republic of Nauru Concerning Operational Cooperation to Suppress Illicit Transnational Maritime Activity (T.I.A.S. 11-908), signed September 8, 2011.

[240] Agreement Between the Government of the United States of America and the Government of the Republic of Palau Concerning Cooperation to Suppress Illicit Activity at Sea (T.I.A.S. 08-320), signed March 20, 2008; Agreement Between the Government of the United States of America and the Government of the Republic of Palau Concerning Operational Cooperation to Suppress Illicit Transnational Maritime Activity (T.I.A.S. 13-0815), signed August 15, 2013; and Stephen Wright, "Palau, United States Expand Maritime Security Arrangements After Chinese Incursions," *BenarNews*, August 30, 2023, https://www.rfa.org/english/news/pacific/palau-us-security-08302023222710.html.

[241] Supplementary Arrangement Between the Government of the United States of America and the Government of the Republic of Panama to the Arrangement Between the Government of the United States of America and the Government of Panama for Support and Assistance from the United States Coast Guard for the National Maritime Service of the Ministry of Government and Justice (T.I.A.S. 02-205.1), signed February 5, 2002.

Papua New Guinea,[242] Republic of Marshall Islands,[243] Samoa,[244] Senegal,[245] Seychelles,[246] Sierra Leone,[247] Tonga,[248] Tuvalu,[249] and Vanuatu.[250] Congress may examine whether sufficient support and resources have been dedicated to enforcement efforts to counter IUU fishing activities, such as capacity-building assistance to coastal nations and joint efforts, including shiprider agreements.

---

[242] Agreement Between the United States of America and the Independent State of Papua New Guinea Concerning Counter Illicit Transnational Maritime Activity Operations (T.I.A.S. 23-816.1), signed May 22, 2023.

[243] Agreement Between the Government of the United States of America and the Government of the Republic of the Marshall Islands Concerning Cooperation in Maritime Surveillance and Interdiction Activities (T.I.A.S. 08-805), signed August 5, 2008 (amended March 19, 2013).

[244] Agreement Between the Government of the United States of America and the Government of the Independent State of Samoa Concerning Operational Cooperation to Suppress Illicit Transnational Maritime Activity (T.I.A.S. 12-602), signed June 2, 2012; and *Maritime Executive*, "Samoa Grants USCG Expanded Enforcement Powers in its EEZ," April 7, 2024, https://maritime-executive.com/article/samoa-grants-uscg-expanded-enforcement-powers-in-its-eez.

[245] Agreement Between the Government of the United States of America and the Government of the Republic of Senegal Concerning Operational Cooperation to Suppress Illicit Transnational Maritime Activity (T.I.A.S. 11-429), signed April 29, 2011.

[246] Agreement Between the Government of the United States of America and the Government of the Republic of Seychelles Concerning Counter Illicit Transnational Maritime Activity Operations (T.I.A.S. 21-727), signed July 27, 2021.

[247] The Sierra Leone agreement with the United States is an *executive agreement*. Agreement Between the Government of the United States of America and the Government of the Republic of Sierra Leone Concerning Cooperation to Suppress Illicit Transnational Maritime Activity, signed June 26, 2009, https://2009-2017.state.gov/documents/organization/153587.pdf. This agreement is an *executive agreement*. Executive agreements are entered into without the advice and consent of the U.S. Senate, but are still binding on the parties under international law. See U.S. Senate, "About Treaties," https://www.senate.gov/about/powers-procedures/treaties.htm.

[248] Agreement Between the Government of the United States of America and the Government of the Kingdom of Tonga Concerning Cooperation in Joint Maritime Surveillance Operations, signed August 24, 2009, https://2009- 2017.state.gov/documents/organization/153588.pdf; and U.S. Department of State, "The United States-Tonga Relationship," July 24, 2023, https://www.state.gov/the-united-states-tonga-relationship/. This agreement is an *executive agreement*. See footnote 247.

[249] Agreement Between the Government of the United States of America and the Government of Tuvalu Concerning Operational Cooperation to Suppress Illicit Transnational Maritime Activity (T.I.A.S. 11-909), signed September 9, 2011; and U.S. Department of State, "U.S. Relations with Tuvalu," June 23, 2022, https://www.state.gov/u-s-relations- with-tuvalu/.

[250] Agreement Between the Government of the United States of America and the Government of the Republic of Vanuatu Concerning Counter Illicit Transnational Maritime Activity Operations (T.I.A.S. 16-1031), signed October 31, 2016.

What Technologies Can Be Used to Identify Vessels Suspected of IUU Fishing?

Earth's vast ocean area enables some fishing fleets to conduct IUU fishing activity unnoticed and presents law enforcement challenges. Technology can play an important role in patrolling the sea for vessels suspected of IUU fishing. Both vessel monitoring systems (VMS) and automatic identification systems (AIS) are widely used to monitor vessel location and movements from remote locations.[251] AIS and VMS are distinct systems that are not interoperable or compatible but may be used in conjunction (Table 3).[252] These systems employ electronic transmitters that can be installed on vessels and send information from ship to ship, ship to shore, or ship to satellite. Data are then relayed to enforcement personnel who monitor information such as vessel identification, date, time, and location. Both VMS and AIS can support law enforcement by allowing patrols to focus on areas with the highest potential for fishing violations, although some stakeholders find AIS to be the better tool for monitoring fishing.[253] The efficacy of these systems depends on whether they are used consistently and provide information on a real-time basis.

Efficacy also depends on whether data such as vessel name, class, flag operator, and owner are available and matched to vessel databases.

Monitoring fishing vessels with AIS to detect illegal fishers may be limited because operators can turn off their systems and "go dark." Some research has found that vessels most often go dark while fishing next to EEZs with contested boundaries, while fishing in EEZs with limited management oversight, and during the transfer of fish between fishing vessels and refrigerated cargo vessels.[254]

---

[251] Global Fishing Watch, "What Is AIS?," https://globalfishingwatch.org/faqs/what-is-ais/; Oceana, "AIS: What Is It?," https://usa.oceana.org/wp-content/uploads/sites/4/4046/oceana_ais_fin_all_hr.pdf (hereinafter Oceana "AIS: What Is It?").

[252] Oceana, "Automatic Identification System," https://usa.oceana.org/wp-content/uploads/sites/4/2023/07/Fact-Sheet-on-AIS-Vessel-Tracking-2023.pdf. Hereinafter Oceana, "Automatic Identification System."

[253] Ibid. Compared with a vessel monitoring system, an automatic identification system has a higher temporal resolution for transmitting signals (i.e., near real-time reporting), a lower cost, and its data are publicly available (see Table 3).

[254] NOAA, NMFS, "Learning More About 'Dark' Fishing Vessels' Activities at Sea," November, 2, 2022, https://www.fisheries.noaa.gov/feature-story/learning-more-about-dark-fishing-vessels-activities-sea.

Table 3. Vessel tracking instruments

	Automatic Identification System (AIS)	Vessel Monitoring System (VMS)
Operational Mode	Provides vessel navigation information (including vessel's identity, type, course, speed, and other safety-related information) in real time, via ship-to-ship, ship-to-shore, ship-to-aircraft, or ship-to-satellite communication	Remotely monitors fishing vessel position in relation to regulatory areas and maritime boundaries, via a scheduled or manual broadcast to satellite receivers and authorized data recipients
Temporal Resolution	Signal transmitted every few seconds	Signal typically transmitted at least once per hour
Approximate Cost	$750-$3,500, no associated fees	$4,000, plus associated fees throughout the vessel's lifetime
Service Provider	Open, nonproprietary	Closed, proprietary protocols
Tamper-Proof	No	Yes
Applicability	Required—per SOLAS V/19 or 33 C.F.R. §164.46—commercial fishing vessels 65 feet long or greater	Required—by NOAA via regulations—for vessels participating in fishing for certain fishery species
Required on Vessels over 65 feet long	Yes	Requirements for VMS use are fishery-specific
Approximate Number of U.S. Vessels	More than 40,000	More than 4,000

Sources: Congressional Research Service, modified from U.S. Coast Guard, "How Does AIS Compare and Contrast with VMS," https://www.navcen.uscg.gov/sites/default/files/pdf/AIS/ Q_AIS_vs_VMS_Comparison_2016.pdf; Oceana, "Automatic Identification System," https://usa.oceana.org/wp- content/uploads/ sites/4/2023/07/Fact-Sheet-on-AIS-Vessel-Tracking-2023.pdf, p. 4; 33 C.F.R. §164.46; 50 C.F.R. §§660.1500-600.1516; National Oceanic and Atmospheric Administration (NOAA), National Marine Fisheries Service (NMFS), "Regional Vessel Monitoring Information," https://www.fisheries.noaa.gov/ national/ enforcement/regional-vessel-monitoring-information; and NOAA, NMFS, "Enforcement: Vessel Monitoring," https://www.fisheries.noaa.gov/topic/ enforcement/vessel-monitoring.

Notes: SOLAS V/19 = Chapter V, Regulation 19, of the International Convention for Safety of Life at Sea. NMFS reports that the U.S. VMS fleet is the largest national VMS fleet in the world.

International and regional organizations, as well as some countries, require the use of AIS on certain vessels. The International Maritime Organization requires large ships, including many commercial fishing vessels, to broadcast their position with AIS.[255] Some RFMOs require the use of VMS, and are considering the utility of AIS, for vessels fishing in their convention

---

[255] International Maritime Organization, "AIS Transponders," https://www.imo.org/en/ OurWork/Safety/Pages/ AIS.aspx.

areas.²⁵⁶ The United States requires its commercial fishing vessels over 65 feet long to have an AIS while operating in U.S. waters.²⁵⁷ Some stakeholders have proposed that the United States apply this requirement to fishing vessels over 49 feet long, which would align with the European Union's requirement for its fishing vessels.²⁵⁸

Some stakeholders are interested in applying machine learning to satellite-based data to improve enforcement patrols for IUU fishing. As part of a worldwide competition, the Defense Innovation Unit, a civilian organization within DOD, and Global Fish Watch solicited developers to apply machine learning to satellite-based synthetic aperture radar (SAR) data to detect vessels that had gone dark.²⁵⁹ SAR technology can penetrate clouds and can be used at night to identify the location and movement of dark vessels. By 2023, the U.S. government had operationalized the machine learning algorithms developed during the competition within the USCG, NOAA, and the U.S. Navy and integrated the model outputs into SeaVision.²⁶⁰ SeaVision, a web-based encrypted sharing network of maritime domain awareness information, uses nonclassified vessel AIS data to display current and past vessel movement within the U.S. EEZ, within the EEZs of partner countries, and on the high seas on a live map.261 SeaVision data may be analyzed to identify illegal fishing activity, among other vessel information.

---

[256] Holly Koehler, *RFMO Vessel Monitoring Systems: A Comparative Analysis to Identify Best Practices*, International Seafood Sustainability Foundation, ISSF Technical Report 2022-06, March 2022.

[257] 33 C.F.R. §164.46(b)(1)(i).

[258] For example, see Oceana, "AIS: What Is It?"; and Center for the Blue Economy, *Turning the Tide: Biden Administration Leadership on Ocean Climate Action & Recommended Next Steps*, June 2024, p. 29.

[259] Defense Innovation Unit, "U.S. Government and Nonprofit Organization Host Prize Competition to Leverage the Latest Technology to Detect and Defeat Illegal Fishing," July 22, 2021, https://www.diu.mil/latest/us-government-and-nonprofit-organization-host-prize-competition-xview3. The USCG, NOAA, and the National Maritime Intelligence-Integration Office also supported the solicitation.

[260] Alex Appel, "AI as a Weapon to Defend the Seas from Illegal Fishing," *Alaska Business*, November 13, 2023, https://www.akbizmag.com/industry/fisheries/illegal-fishing-ai/. For more information about SeaVision, see the "What Actions Are U.S. Agencies Taking to Address IUU Fishing?" section of this report.

Chapter 3

Combating Illegal Fishing: Clear Authority Could Enhance U.S. Efforts to Partner with Other Nations at Sea[*]

United States Government Accountability Office

Abbreviations

AFRICOM	U.S. Africa Command
AIS	Automatic identification system
DOD	Department of Defense
EEZ	Exclusive economic zone
FAO	United Nations Food and Agriculture Organization
IUU	Illegal, unreported, and unregulated
Maritime SAFE Act	Maritime Security and Fisheries Enforcement Act
MOTR	Maritime Operational Threat Response
NMIO	National Maritime Intelligence-Integration Office
NOAA	National Oceanic and Atmospheric Administration
RFMO	Regional fisheries management organization
UNCLOS	United Nations Convention on the Law of the Sea
VIIRS	Visible Infrared Imaging Radiometer Suite
VMS	Vessel monitoring system

---

[*] This is an edited, reformatted and augmented version of the United States Government Accountability Office Report to Congressional Requesters, Publication No. GAO-22-104234, dated November 2021.

In: Ongoing Efforts to Combat Illegal, Unreported …
Editor: Gordon B. Maddox
ISBN: 979-8-89530-858-5
© 2026 Nova Science Publishers, Inc.

United States Government Accountability Office

Why GAO Did This Study

IUU fishing undermines the economic and environmental sustainability of the fishing industry in the U.S. and globally. IUU fishing encompasses many illicit activities, including under- reporting the number of fish caught and using prohibited fishing gear.

While the illicit nature of IUU fishing means its consequences can only be estimated, a recent study estimates catches from IUU fishing could cause global economic losses up to $50 billion annually. A variety of federal agencies coordinate with one another, as well as internationally, to address IUU fishing at sea.

GAO was asked to review federal efforts to combat IUU fishing outside of U.S. waters. This report examines how the U.S. (1) works with other nations to address IUU fishing at sea, (2) identifies potential incidents of IUU fishing at sea, and (3) coordinates its interagency efforts to combat IUU fishing at sea and the extent to which selected efforts are consistent with leading collaboration practices. GAO reviewed various international agreements and the mechanisms that support these efforts, as well as other relevant agency documents. We also spoke with officials from the U.S. Coast Guard, NOAA, and DOD, among others, about their approaches to identifying and combating IUU fishing at sea.

What GAO Recommends

GAO is making one recommendation to the Department of Defense to determine whether it has the authority to continue Operation Junction Rain and, if not, seek the authority to do so. DOD partially concurred with our recommendation.

What GAO Found

The U.S. works with other nations through multilateral agreements to collectively manage high seas fisheries. For example, the U.S. is a member of nine regional fisheries management organizations (RFMO), which are treaty-based organizations of nations with an interest in managing and conserving fisheries in specific regions of the sea. These organizations establish rules for vessels fishing in the RFMO agreement area, such as limits on the numbers and types of fish that can be caught. In addition, the U.S. establishes bilateral agreements and conducts at-sea operations focused on strengthening other nations' capacity to manage their own fisheries and fleets. For example, the Department of Defense (DOD) leads a program aimed at building African partner nations' capability to enhance maritime security and enforce their maritime laws. However, DOD officials told us that, as a result of changes to the 2017 National Defense Authorization Act, the department no longer has clear authority to conduct the operational phase of this program—known as Operation Junction Rain. By determining whether it has the authority to conduct this operation, and, if not, seeking such authority, DOD could continue efforts to support African partner nations' capability to enforce fisheries laws and regulations, which in turn helps them work to counter illegal, unreported, and unregulated (IUU) fishing.

Coast Guard Officials Preparing to Board and Inspect a Fishing Vessel

Source: State Deartment and U.S. Coast Guard. |GAO-22-104234.

The U.S. collects and analyzes information from various sources to identify potential IUU fishing at sea outside of U.S. waters. For example, Coast Guard analyzes vessel location data to identify movements that may signal potential IUU fishing, and officials told us they use this data analysis to help to guide at- sea patrol operations to target these vessels.

Several interagency groups and processes help coordinate federal efforts to combat IUU fishing at sea. For example, an interagency working group, established by the Maritime Security and Fisheries Enforcement Act in 2019, coordinates U.S. efforts to address IUU fishing government-wide. We found that the working group generally followed selected leading collaboration practices, such as developing a written work plan. The working group's tasks include assessing areas for increased agency information-sharing on IUU fishing-related matters, identifying priority regions and nations, and developing a 5-year strategic plan to combat IUU fishing and enhance maritime security.

November 5, 2021

The Honorable Roger F. Wicker
Ranking Member
Committee on Commerce, Science, and Transportation
United States Senate

The Honorable Dan Sullivan
Ranking Member
Subcommittee on Oceans, Fisheries, Climate Change, and Manufacturing
Committee on Commerce, Science, and Transportation
United States Senate

The Honorable John Thune United States Senate

Illegal, unreported, and unregulated (IUU) fishing undermines the economic and environmental sustainability of fisheries and fish stocks, both in the U.S. and globally.[1] Potential effects of IUU fishing also include jeopardizing food and economic security and benefitting transnational crime by supporting illicit networks, such as narcotics trafficking or other criminal activities at sea. IUU fishing encompasses many illicit activities, ranging from underreporting the number and types of fish caught to using prohibited fishing gear, such as illegal driftnets.[2] The illicit nature of IUU fishing means that the size of the problem and its negative consequences can only be roughly estimated; however, according to estimates in a recent study, global illicit trade in catches

---

[1] A fishery refers to one or more stocks of fish that can be treated as a unit for conservation and management purposes and that are identified on the basis of geographical, scientific, technical, recreational, and economic characteristics. 16 U.S.C. § 1802(13)(A). A stock of fish refers to a species, subspecies, geographical grouping, or other category of fish capable of being managed as a unit. *Id.* § 1802(42). A stock of fish may be one species or a group of comparable species.

[2] Driftnet fishing is a method using large nets that are allowed to drift with the current and that are designed to entangle fish in the net's webbing.

from IUU fishing causes losses of up to 50 billion dollars annually from legitimate markets.[3]

The U.S. is one of many nations working to combat IUU fishing, and a variety of federal agencies are involved in U.S. efforts to address this global issue. For example, the U.S. Coast Guard, which is a component of the Department of Homeland Security, is the lead agency for at-sea law enforcement. The Department of Defense (DOD) engages in at-sea exercises with other countries to help partner nations build maritime security capacity, which officials say can contribute to combating IUU fishing. The National Oceanic and Atmospheric Administration (NOAA), within the Department of Commerce, has subject matter expertise in fisheries management and fisheries law enforcement. Within NOAA, the Office of General Counsel and the National Marine Fisheries Service's Office of Law Enforcement and Office of International Affairs and Seafood Inspection have roles in combating IUU fishing at sea, according to NOAA officials. The State Department has a role in negotiating and implementing international treaties and agreements that address IUU fishing in coordination with other nations. The National Maritime Intelligence-Integration Office (NMIO) facilitates information sharing and collaboration across the Global Maritime Community of Interest, which consists of federal, state, local, tribal, and territorial governments; the maritime industry; academia; and international partners.[4]

You asked us to review federal efforts to combat IUU fishing outside of U.S. waters.[5] This report examines how the U.S. (1) works with other nations to address IUU fishing at sea, (2) identifies potential incidents of IUU fishing at sea, and (3) coordinates its interagency efforts to combat IUU fishing at sea and the extent to which selected efforts are consistent with leading collaboration practices.

To describe how the U.S. works with other nations to address IUU fishing, we reviewed various multi- and bilateral agreements the U.S. has established with other nations, as well as summaries of those agreements. For example, we reviewed collective agreements to which the U.S. is a member, such as

---

[3] U.R. Sumaila, D. Zeller, L. Hood, M.L.D. Palomares, Y. Li, and D. Pauly, "Illicit Trade in Marine Fish Catch and Its Effects on Ecosystems and People Worldwide," *Science Advances*, vol. 6, no. 9 (2020).

[4] The Office of the Director of National Intelligence, in cooperation with the Navy and the Coast Guard, created NMIO in 2009 to advance governmental collaboration and unity of effort.

[5] Our report focuses on operations at sea, but excludes efforts within U.S. territorial waters, which generally extend up to 12 nautical miles from the coastline, as well as the U.S. exclusive economic zone, which generally extends up to 200 nautical miles from the coastline.

agreements establishing regional fisheries management organizations (RFMO), and other multilateral agreements, such as the Port State Measures Agreement. We also reviewed bilateral shiprider agreements.[6] Additionally, we interviewed relevant agency officials about the development and implementation of these agreements, including officials from the Coast Guard, NOAA, State Department, DOD, and NMIO.

To describe how the U.S. identifies potential incidents of IUU fishing at sea, outside of U.S. waters, we reviewed documents and interviewed agency officials about the information used to identify potential IUU fishing at sea and how it is obtained. For example, we discussed the collection and use of vessel location data, as well as the planning and operation of Coast Guard at-sea patrols. We also spoke with Coast Guard officials about U.S. boardings and inspections of foreign-flagged vessels outside of U.S. waters.

To identify and describe how the U.S. coordinates its interagency efforts to combat IUU fishing at sea, we reviewed documents and interviewed officials about interagency collaboration through the Maritime Security and Fisheries Enforcement (SAFE) Act IUU fishing working group and other interagency working groups, including the International Maritime Domain Awareness Working Group.[7] For the Maritime SAFE Act working group, we reviewed documents, including a work plan and meeting summaries. To discuss the early operations of the group since its inception in December 2019, we also spoke with officials from NOAA, which chairs the group; the Coast Guard; State Department; DOD; and NMIO. We compared the interagency collaboration to date with selected leading practices for interagency collaboration identified in our past work.[8] We focused on applying those leading practices to the Maritime SAFE Act working group because it is designed to facilitate a government-wide approach to addressing IUU fishing. We selected five of the seven leading collaboration practices because they were the most relevant to the working group. The five leading practices we selected were: (1) identifying and sustaining leadership, (2) including relevant

---

[6] Bilateral shiprider agreements are agreements which allow a law enforcement official from one party to embark on a law enforcement vessel of the other party. These agreements are designed to help partner nations enforce fisheries law and to prepare personnel from those nations for independent enforcement of fisheries law in the long term.

[7] The Maritime SAFE Act was enacted as part of the National Defense Authorization Act for Fiscal Year 2020. Pub. L. No. 116-92, div. C, tit. XXXV, subtit. C, 133 Stat. 1997 (2019). Among other things, the SAFE Act established a collaborative interagency working group on maritime security and IUU fishing. *Id.* § 3531.

[8] GAO, *Managing for Results: Key Considerations for Implementing Interagency Collaborative Mechanisms*, GAO-12-1022 (Washington, D.C.: Sept. 27, 2012).

parties, (3) developing and updating written guidance and agreements, (4) clarifying roles and responsibilities, and (5) defining outcomes and monitoring accountability.

We conducted this performance audit from April 2020 to September 2021 in accordance with generally accepted government auditing standards.

Those standards require that we plan and perform the audit to obtain sufficient, appropriate evidence to provide a reasonable basis for our findings and conclusions based on our audit objectives. We believe that the evidence obtained provides a reasonable basis for our findings and conclusions based on our audit objectives.

Background

Maritime Zones

Applicable requirements for fishing vessels at sea, and thus the kinds of fishing that are permissible, vary depending on the maritime zone. A nation's territorial waters are generally a zone extending from a nation's coastline up to 12 nautical miles away;[9] coastal nations have sovereignty and jurisdiction over this zone. Beyond and adjacent to the territorial sea, coastal nations generally have an exclusive economic zone (EEZ) up to 200 nautical miles from their coastlines.[10] In this zone, a coastal nation has certain rights, including sovereign rights for the purpose of exploring and exploiting, conserving, and managing the natural resources of the waters.

Beyond EEZs, the ocean is generally defined as "high seas" and is considered international waters. According to a report from the nongovernmental organization Pew Charitable Trusts, the high seas represent approximately two-thirds of the world's oceans. Figure 1 illustrates the locations of maritime zones.

---

[9] Territorial sea begins at a nation's baseline, which is defined in the 1982 United Nations Convention on the Law of the Sea (UNCLOS) as generally the low-water line along the coast as marked on large-scale charts officially recognized by the coastal nation. For the purposes of this report, we call this a nation's coastline.

[10] International law governing exclusive economic zones (EEZ) was established by UNCLOS. The U.S. is not a party to UNCLOS; however, according to officials from NOAA and the State Department, the U.S. recognizes that UNCLOS reflects customary international law.

Source: GAO analysis of United Nations Convention on the Law of the Sea. | GAO-22-104234.

Figure 1. Maritime Zones of the Oceans.

On the high seas, vessels are subject to the laws of their flag state.[11] Additionally, because flag states commit to implementing international agreements and conventions to which they are a member, a nation's vessels are subject to applicable rules established by international agreements and conventions, as implemented by the nation. These agreements and conventions, including RFMO agreements, generally cover specific geographic regions of the high seas. RFMOs are treaty- based international bodies comprising nations that share an interest in managing and conserving fisheries in specific regions of the high seas.[12] RFMOs establish binding conservation measures to manage and conserve particular species of fish or other living marine resources within specific geographic regions of the oceans. For example, the International Commission for the Conservation of Atlantic Tunas RFMO agreement establishes a limit on allowable catch of certain species of tuna within the agreement area. Violations of RFMO conservation measures are generally considered IUU fishing.

---

[11] The flag state of a vessel is the nation of jurisdiction under whose laws the vessel is registered or licensed and is deemed the nationality of the vessel.

[12] The U.S. belongs to nine RFMOs where the U.S. is a coastal nation or has a fishing interest, according to State Department officials: (1) North Pacific Fisheries Commission, (2) Northwest Atlantic Fisheries Organization, (3) International Commission for the Conservation of Atlantic Tunas, (4) Western and Central Pacific Fishery Commission, (5) Inter-American Tropical Tuna Commission, (6) Commission for the Conservation of Antarctic Marine Living Resources, (7) South Pacific Regional Fishery Management Organization, (8) The North Atlantic Salmon Conservation Organization and (9) the North Pacific Anadromous Fish Commission. Additionally, the U.S. acts as an observer to other RFMOs covering fisheries that are harvested for U.S. import. There are also a number of other international fishing agreements that function similarly but that are not considered RFMOs by the State Department. According to a report from the nongovernmental organization Pew Charitable Trusts, there are approximately 17 RFMOs worldwide.

Definition of IUU Fishing

IUU fishing is a broad term that generally includes activities that violate national law or international fishing regulations or agreements.[13] NOAA defines each aspect of IUU fishing as follows:

- Illegal fishing refers to fishing activities conducted in contravention of applicable laws and regulations, including those adopted at the regional and international level.
- Unreported fishing refers to fishing activities that are not reported or are misreported to relevant authorities in contravention of national laws and regulations or reporting procedures of a relevant RFMO.
- Unregulated fishing occurs in geographic areas or for specific species of fish for which there are no applicable conservation or management measures, and when fishing activities are conducted in a manner inconsistent with a nation's responsibilities for the conservation of living marine resources under international law. Fishing activities are also unregulated when occurring in an RFMO- managed area and conducted by vessels without nationality, or by those flying a flag of a nation or fishing entity that is not party to the RFMO, in a manner inconsistent with the conservation measures of that RFMO.[14]

IUU fishing encompasses many illicit activities that can occur both within a nation's EEZ as well as on the high seas. For example, within national EEZs, vessels may engage in IUU fishing by fishing without an appropriate license or fishing above a nationally established quota. On the high seas, examples of IUU fishing include fishing out of season or fishing in a prohibited area. Figure 2 below illustrates common types of IUU fishing.

---

[13] Although human trafficking and forced labor are illegal under U.S. law, they are not generally included in regulatory definitions of IUU fishing, including the definition under the High Seas Driftnet Moratorium Protection Act or the definition in the Maritime SAFE Act. For the purposes of this report, IUU fishing does not include fishing with forced labor.

[14] Under the Maritime SAFE Act, the term IUU fishing means illegal fishing, unreported fishing, or unregulated fishing as such terms are defined in paragraph 3 of the International Plan of Action to Prevent, Deter, and Eliminate Illegal, Unreported and Unregulated Fishing, adopted at the 24th Session of the Committee on Fisheries in Rome on March 2, 2001. NOAA's definitions, above, are also based on the definitions in the International Plan of Action.

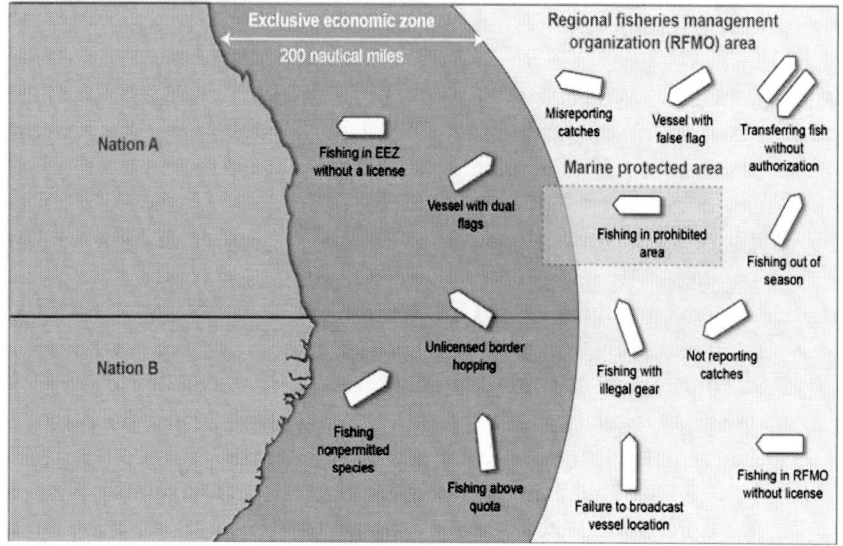

Source: GAO analysis of agency information. | GAO-22-104234.

Figure 2. Common Types of Illegal, Unreported, and Unregulated Fishing.

U.S. Laws and Federal Agencies

The U.S. works through international relationships to combat IUU fishing on the high seas. In doing so, U.S. jurisdiction is generally limited to its own flagged vessels; however, some domestic laws are relevant to U.S. efforts to combat IUU fishing by foreign-flagged vessels operating outside of U.S. waters, including the following:

- The *Maritime SAFE Act* was enacted in 2019 for several purposes, including to support a government-wide approach to counter IUU fishing and related threats to maritime security; improve data sharing that enhances surveillance, enforcement, and prosecution against IUU fishing and related activities at a global level; and support coordination and collaboration to counter IUU fishing internationally in priority regions.[15] The act established, among other things, an interagency working group on maritime security and IUU fishing. For

---

[15] Pub. L. No. 116-92, div. C, tit. XXXV, subtit. C, §§ 3531-3572, 133 Stat. 1997 (2019) (codified at 16 U.S.C. §§ 8001-8041).

the purposes of this report, we refer to this group as the Maritime SAFE Act working group.
- The *High Seas Fishing Driftnet Moratorium Protection Act* prohibits the U.S. from entering into international agreements that would prevent full implementation of the UN Moratorium on Large- Scale High Seas Driftnets.[16] The act, and the statutes implementing the various RFMO conventions that the United States is party to, apply to U.S. vessels, but the U.S. can also apply those laws to stateless vessels operating on the high seas as if they were U.S. vessels.
- The *Magnuson-Stevens Fishery Conservation and Management Reauthorization Act of 2006*, which amended the High Seas Driftnet Fishing Moratorium Protection Act, established a process for identifying nations for IUU fishing in a biennial report to Congress.[17]

Efforts to combat IUU fishing on the high seas require coordination and information sharing among a number of federal agencies that have different roles and responsibilities (See Table 1).

Table 1. U.S. Agency Roles in International Efforts to Combat IUU fishing

Agency	Key roles and responsibilities
National Oceanic and Atmospheric Administration (NOAA), within the Department of Commerce	NOAA has subject matter expertise on fisheries management and illegal, unreported, and unregulated (IUU) fishing; accordingly, it consults with the State Department in negotiations of some international agreements, and it represents U.S. efforts in some multilateral agreements that manage high seas and shared fisheries. NOAA also works with other nations to strengthen enforcement and data collection programs and participates in some at-sea exercises with such partner nations. Additionally, according to NOAA officials, the agency can enforce U.S. marine resource laws, including those targeting IUU fishing and trafficking in IUU fish and fish products.

---

[16] Pub. L. No. 104-43, tit. VI, 109 Stat. 391 (1995) (codified as amended at 16 U.S.C. §§ 1826d-1826k).

[17] Pub. L. No. 109-479, § 403(a), 120 Stat. 3575, 3626 (codified as amended at 16 U.S.C.§ 1826h). For the biennial reports produced since 2009, see https://www.fisheries.noaa.gov/international/report-iuu-fishing-bycatch-and-shark-catch. Once a nation or entity is identified in the biennial report, NOAA enters into a two-year consultation period to press for necessary measures to address the issue for which it was identified. Following these consultations, NOAA determines whether to negatively or positively certify the identified nation or entity in the next report to Congress. A positive certification is issued if the nation has provided evidence of actions that address the activities for which it was identified. A negative certification may result in denial of U.S. port access for fishing vessels of that nation and potential import restrictions on fish or fish products.

Table 1. (Continued)

Agency	Key roles and responsibilities
U.S. Coast Guard	The Coast Guard generally serves as the lead agency for at-sea enforcement of international fisheries agreements—including boarding and inspecting vessels suspected of IUU fishing, which it can do under the authority of some regional fisheries management organizations—and for identifying potential IUU fishing at sea. The Coast Guard also works closely with other nations in efforts to build capacity to manage their own fisheries and fleets.
State Department	The State Department leads diplomatic efforts, including negotiating new RFMOs and other international agreements, and maintains contact with other nations during implementation of these agreements. It also coordinates with other nations through diplomatic channels in cases where vessels flagged under those nations are identified as engaging in IUU fishing and a government- to-government approach to addressing the potential violation is necessary.
Department of Defense (DOD)	DOD engages in periodic at-sea exercises with other countries to help build other nations' maritime security. It generally does so though its geographic combatant commands. According to DOD officials, these exercises build partner nations' maritime security capacity, which can help them to manage their own fisheries and fleets.
Navy	The Navy contributes expertise, personnel, and resources for training and building partner nations' capacity to provide maritime security. According to DOD officials, such capacity can help partner nations manage their own fisheries.
National Maritime Intelligence- Integration Office (NMIO)	Established by the Office of the Director of National Intelligence, which is administered by the Navy, NMIO facilitates maritime information sharing within the government and provides unified intelligence support to U.S. policymakers.

Source: GAO analysis of agency information. | GAO-22-104234.

U.S. Works with Other Nations to Address IUU Fishing through Agreements and Capacity-Building Efforts, but DOD Lacks Clarity on Its Authority to Fully Execute Some Efforts

The U.S. works with other nations through multilateral agreements to develop measures to collectively manage high seas fisheries. In addition, the U.S. establishes bilateral agreements focused on building other nations' capacity to manage their own fisheries and fleets, including DOD exercises with partner nations designed to increase other nations' maritime security, which can include building fisheries law enforcement capacity. However, DOD lacks clarity on whether it has all the necessary authority to conduct parts of its capacity-building efforts in Africa.

U.S. Works with Other Nations through Multilateral Agreements for Collective Management of High Seas Fisheries

The U.S. is a member of various multilateral agreements with other nations to collectively and sustainably manage high seas fisheries and combat IUU fishing. Through RFMO agreements, the U.S. works collectively with other member nations to patrol areas of the high seas covered by RFMOs to identify potential IUU fishing. For example, Operation North Pacific Guard is an IUU fishing-focused operation by Japan, China, Russia, South Korea, Canada, and the U.S. that patrols areas of the northern Pacific Ocean covered by three RFMO agreements.[18] In 2019, this operation identified 58 violations of RFMO conservation measures. The vessels were reported to the relevant RFMOs and their individual flag states for further action. See figure 3 below for photos of a Coast Guard vessel interdicting a vessel using an illegal high seas driftnet, and Coast Guard officials preparing to conduct a law enforcement boarding under measures established by the Western and Central Pacific Fisheries Commission RFMO.

Source: State Department and U.S. Coast Guard. | GAO-22-104234.

Figure 3. Coast Guard Vessel Interdicting Illegal Fishing and Officials Preparing to Conduct a Law Enforcement Boarding.

The U.S. and other RFMO member nations collaborate to ensure violations of RFMO conservation measures are addressed. When U.S. officials identify potential violations, they work with other RFMO member nations and relevant flag states to take appropriate actions. For example, in 2020, the Coast Guard identified a vessel violating the International Commission for the Conservation of Atlantic Tunas conservation measures by fishing without

---

[18] The operation patrols areas covered by the North Pacific Fisheries Commission, Western and Central Pacific Fisheries Commission, and North Pacific Anadromous Fisheries Commission RFMO agreements.

registration in waters covered by the RFMO. Moreover, the Coast Guard determined the vessel was not legally flagged and was therefore presumed to be a stateless vessel. NOAA prepared a correspondence to the Secretariat of the RFMO advising the Secretariat of the sighting and flagless status of the vessel. This message to the Secretariat requested the reported information be forwarded to all member states of the RFMO. Member states that received the message would be on the lookout for the vessel for tracking purposes and be able to board and inspect if it came into their ports, according to NOAA officials.

In addition to RFMOs, the U.S. participates in other multilateral agreements and coalitions to collectively manage high seas fisheries and address IUU fishing on the high seas. Other multilateral agreements in which the U.S. participates include the following:

- The Port State Measures Agreement – This agreement is the first binding international agreement to specifically target IUU fishing according to the United Nations Food and Agriculture Organization.[19] U.S. implementation of the agreement includes sharing vessel inspection information with other nations, organizations, and RFMOs. The agreement seeks to block fishery products derived from IUU fishing from reaching markets by encouraging communication among nations on what vessels are known to have engaged in IUU fishing on the high seas and denying them port access. The agreement also provides a framework through which the U.S. works with other nations to help achieve United Nations sustainable development goals, one of which was to end IUU fishing worldwide by 2020.[20] In support of this agreement, NOAA developed an international training program to provide technical assistance to global partners working to implement the agreement. NOAA officials said they conducted this

---

[19] The United Nations Food and Agriculture Organization (FAO) is the specialized agency of the UN that leads international efforts to defeat hunger and plays a lead role in supporting nations in achieving the 17 sustainable development goals established in 2015. FAO approved the Agreement on Port State Measures to Prevent, Deter and Eliminate Illegal, Unreported and Unregulated Fishing, which entered into force on June 5, 2016. Sixty-nine nations, including the U.S., have become parties to the agreement.

[20] While the 2020 deadline originally stated in the sustainable development goals has passed, FAO reported in December 2020 that member nations have developed frameworks of binding and voluntary international instruments, which, if fulfilled, will lead to preventing, deterring and eliminating IUU fishing. Food and Agriculture Organization of the United Nations, *Combatting Illegal, Unreported and Unregulated Fishing*, 34th session (Rome, Italy: February 2021), 3.

program in Southeast Asia and South America, with financial and organizational support from the State Department and the U.S. Agency for International Development.

- *Quadrilateral Defense Coordination Group.* This multilateral security coordination group collaboratively identifies potential IUU fishing in areas outside of U.S. waters. The group brings together defense and security agencies from Australia, France, New Zealand, and the U.S. to coordinate maritime surveillance to reduce IUU fishing in Pacific Island countries' EEZs and in adjacent areas of the high seas. For example, in 2018, the group conducted a large-scale surveillance operation in support of the Pacific Islands Forum Fisheries Agency.[21] Member nations contributed resources to surface and aerial surveillance. With over 250 personnel involved and 14.1 million square kilometers of ocean included, it was one of the biggest fisheries surveillance operations on record, according to agency officials. During the operation, 257 vessels suspected of IUU fishing were detected, and crew of 177 vessels were interrogated at sea.

- *Agreement to Prevent Unregulated High Seas Fisheries in the Central Arctic Ocean.* This multilateral agreement, to which the U.S. is a party, went into effect in June 2021 and prohibits unregulated fishing in the high seas of the Central Arctic Ocean for 16 years. This area borders the EEZs of the U.S. and other Central Arctic coastal nations. The agreement recognizes that large portions of the central Arctic Ocean were previously covered by ice, limiting vessel access; however, the ice has diminished in recent years, providing new access to vessels. The agreement establishes a joint program of scientific research and monitoring to gain a better understanding of Arctic Ocean ecosystems and provides that commercial fishing in the region will not be authorized until international mechanisms are in place to ensure its sustainability. This agreement is the first multilateral agreement of its kind to take a binding approach to protecting an area from commercial fishing before such fishing has begun, according to the State Department.

---

[21] The Pacific Islands Forum Fisheries Agency is an intergovernmental agency that facilitates cooperation and coordination on fishery policies among its member states to conserve migratory tuna stocks, for the benefit of the peoples of the region.

U.S. Works with Other Nations to Help Build Capacity for Managing Their Fisheries and Fleets, but DOD Lacks Clarity on Its Authority to Continue an Operation That Helped African Nations

As part of its efforts to address IUU fishing, the U.S. works with other nations to help them build capacity to manage their own fisheries and fleets, enabling them to better establish and enforce their fisheries law over their flagged vessels. Examples of how the U.S. partners with nations to build capacity include bilateral shiprider agreements; various efforts managed by NOAA, with cooperation from other agencies; and DOD-led exercises with other nations.

The goals of bilateral shiprider agreements are to help partner nations enforce their fisheries law and to prepare personnel from those nations for independent enforcement of fisheries law in the long term. The U.S. has entered into 15 shiprider agreements that address IUU fishing.[22] Such agreements generally allow the Coast Guard to collaborate with other nations by

- partnering with foreign personnel for training,
- exchanging information,
- allowing partner personnel aboard U.S. vessels to exercise boarding and inspection provisions within their EEZs, and
- taking actions against potential violations by vessels flagged to partner nations on the high seas.

Through operation of shiprider agreements between 2016 and 2020, the Coast Guard boarded and inspected 199 fishing vessels in cooperation with partner nations. Those inspections discovered 25 IUU fishing violations, according to information from Coast Guard officials.[23]

U.S. agencies are working to create new shiprider agreements and expand existing shiprider agreements to include IUU fishing enforcement provisions, according to Coast Guard officials. Those officials noted that these efforts are

---

[22] According to the State Department, the U.S. has shiprider agreements that address IUU fishing with the following nations: Cape Verde, Cook Islands, Federated States of Micronesia, Fiji, Kiribati, Nauru, Palau, Republic of Marshall Islands, Samoa, Senegal, Sierra Leone, The Gambia, Tonga, Tuvalu, , and Vanuatu.

[23] Coast Guard officials told us they compiled these data from after-action reports that their law enforcement staff review, validate, and assess on a quarterly basis as part of the Law Enforcement Planning and Assessment System.

partly in response to the Maritime SAFE Act, which calls for including counter-IUU fishing provisions in existing shiprider agreements in which the U.S. is a party, and entering into new shiprider agreements that include counter-IUU fishing provisions with priority flag states and nations in priority regions.[24]

In addition to Coast Guard collaboration efforts, NOAA develops capacity- building partnerships with other nations, and NOAA officials told us these partnerships enhance partner nations' efforts to combat IUU fishing, including through effective investigation and prosecution of fisheries cases. Specifically, NOAA is authorized, including through U.S. implementation of the Port State Measures Agreement, to provide assistance to other nations to strengthen their efforts to combat IUU fishing through training, technical, and legal assistance. Accordingly, NOAA has taken actions such as providing counter-IUU fishing technical assistance and training for partner nations in Southeast Asia, Africa, Latin America, and the Caribbean. NOAA officials told us these activities can enhance monitoring, control, and surveillance capacities in partner nations, and help partner nations strengthen their fisheries management laws and regulations.

The U.S. also partners with other nations through DOD-led at-sea exercises designed to help them build maritime security capacity, which can contribute to partner nation capacity to address IUU fishing in their territorial waters as well as IUU fishing committed by their flagged vessels on the high seas. DOD's geographic combatant commands lead these exercises, with support from other federal agencies, including NOAA and the Coast Guard.[25] For example, through the U.S. Africa Command's (AFRICOM) African Maritime Law Enforcement Partnership program, the Coast Guard and DOD work together to build African partner nations' capability to enhance maritime security and enforce their maritime laws at sea through real-world combined maritime law enforcement operations.

This program, which started in 2008, has several phases that include risk assessments, classroom training, and joint exercises, and culminate with

---

[24] See 16 U.S.C. § 8014(a). The Maritime SAFE Act working group is currently working to define and identify the priority flag states and regions, according to NOAA officials. Officials from the Department of State and the Coast Guard told us that the U.S. is currently negotiating or renegotiating shiprider agreements with a number of nations, and that those nations will not be publicly identified until those agreements have been finalized.

[25] The Department of Defense has 11 unified combatant commands, each of which has a geographic or functional mission that provides command and control of military forces in peace and war.

incremental U.S. withdrawal at the conclusion of each year's operation, according to DOD officials.

As part of the program, the U.S. previously worked with African partner nations under Operation Junction Rain. Through this operation, Coast Guard law enforcement personnel embarked with partner nation personnel aboard a U.S. Naval ship, Coast Guard ship, or African partner nation vessel to patrol the African partner nation's EEZ and assist the partner nation in enforcing their fishery laws and regulations. However, the 2017 National Defense Authorization Act consolidated a number of authorities for payment of personnel expenses related to certain security cooperation with foreign governments, including the authority on which DOD had previously relied to conduct Operation Junction Rain.[26] As a result of this legislative change, in 2019, legal counsel for AFRICOM and the Office of the Secretary of Defense determined that AFRICOM no longer had the authority to expend funds needed to conduct this operation or any other maritime law enforcement operations in the region, according to AFRICOM officials. Consequently, the final Operation Junction Rain took place in 2019, according to DOD officials. In 2020, AFRICOM officials submitted a legislative proposal to the Office of the Secretary of Defense to request the authority to conduct Operation Junction Rain.

However, these officials told us that the proposal was ultimately withdrawn from consideration, and that officials were exploring other potential authorities. DOD officials told us that, as of June 2021, they still do not believe they have sufficient authority to conduct this operation.

According to the Maritime SAFE Act, it is the policy of the U.S. to, among other things, (1) develop holistic diplomatic, military, law enforcement, economic, and capacity-building tools to counter IUU fishing, and (2) promote global maritime security through improved capacity and technological assistance to support improved maritime domain awareness. Further, according to a DOD document, the ability of African partner nations to enforce their laws at sea directly affects their economic and food security and national stability, and the safety and stability of coastal African nations has significant impacts on U.S. national security and those of U.S. partners and allies. According to DOD officials, Operation Junction Rain yielded significant positive results in developing African partner nations' capacity to strengthen fisheries law enforcement along their nearly 19,000 miles of coastline. If DOD determines whether it has the authority to conduct Operation Junction Rain,

---

[26] Pub. L. No. 114-328, § 1243, 130 Stat. 2000, 2514 (2016).

DOD could either resume the program or seek the requisite authority to do so. Officials told us resuming the program would support African partner nations in developing their ability to enforce fisheries laws and regulations, which in turn would help them work to counter IUU fishing both in their EEZs and on the high seas when committed by their flagged vessels.

U.S. Leverages Various Information Sources to Identify Potential IUU Fishing At Sea

The U.S. collects and analyzes information from various sources to identify potential IUU fishing at sea outside of U.S. waters. For example, technology for tracking vessel location at sea helps U.S. agencies identify movements of fishing vessels on the high seas that may indicate potential IUU fishing.[27] In addition, Coast Guard at-sea operations—which include patrols, boardings, and inspections under the authority of RFMOs—can lead to identification of IUU fishing.

Federal Agencies Use Vessel Location Data to Help Identify Indicators of IUU Fishing at Sea

Federal agencies analyze vessel location data obtained through a variety of tracking technologies and other sources to help identify potential incidents of IUU fishing (see sidebar). Coast Guard officials told us U.S. government agencies—such as the Coast Guard and NOAA—directly collect some vessel location data and that they also receive some vessel location data from other nations. For example, in 2018 during the Coast Guard's annual fisheries enforcement operation, the Department of Fisheries and Oceans of Canada coordinated on the use of radar satellite imagery that was instrumental in identifying IUU fishing in the northern Pacific Ocean. Additionally, Coast Guard officials said they use international data provided through the nonprofit Global Fishing Watch, which has access to six other nations' location data and

---

[27] Some RFMOs that the U.S. is a member of have adopted measures requiring vessels to report location through a satellite-based monitoring system; however, the specifics of the broadcasting requirements vary and, according to NOAA officials, access to these data may be limited to the flag state in some cases.

satellite location data from the European Space Agency, according to Global Fishing Watch officials.[28]

NOAA and the Coast Guard analyze the data to identify vessel movements and actions that may indicate IUU fishing. Such movements and actions include entering another nation's EEZ, moving in patterns that signal illegal transshipping,[29] and ceasing to transmit positional data, according to Coast Guard officials. The Coast Guard also partners with Global Fishing Watch for data analysis, according to Coast Guard officials.

Coast Guard officials told us they use this data analysis to develop lists of vessels suspected of IUU fishing, which help guide at-sea patrol operations to target these vessels. Coast Guard officials told us the lists of suspected vessels contribute to successful identification of IUU fishing at sea. For example, the Coast Guard identified and interdicted a listed vessel in 2018, according to officials. The Coast Guard included this vessel on a list of suspected vessels after analysis of location data indicated the vessel was moving in ways characteristic of using prohibited high seas driftnets.[30] Coast Guard officials told us images of the vessel captured by Coast Guard aircraft confirmed this suspicion. Coast Guard officials told us they intercepted the vessel, confirmed illegal use of driftnets, and escorted the vessel to authorities of its flag state. According to Coast Guard officials, authorities of the flag state prosecuted the case, imprisoned several people, and destroyed the vessel.

---

[28] Global Fishing Watch's mission is to advance ocean governance through increased transparency of human activity at sea. In support of this, the organization creates and shares maps, data, and analysis tools. The organization signed a memorandum of understanding with the Coast Guard Research and Development Center in 2019 that formalizes coordination to help deter IUU fishing, among other purposes. Representatives from Global Fishing Watch told us they also collaborate and share information with NOAA.

[29] For the purposes of this report, transshipping refers to the transfer of fish or other goods from one vessel to another at sea.

[30] High seas driftnet fishing involves deploying large mesh curtains up to 10 miles wide and approximately 50 feet deep to trap catch behind the gills. Driftnets catch both target and non-target marine species and can result in overharvesting. The United Nations General Assembly adopted a resolution in 1991 calling for a worldwide moratorium on all large-scale pelagic driftnet fishing, and the U.S. implemented this resolution through the High Seas Driftnet Fisheries Enforcement Act in 1992.

> ### Examples of Vessel Location Tracking Technologies
>
> *Automatic Identification System (AIS)* is a shipboard broadcast system that uses radio waves to continuously send and receive location and position information within approximately 20 nautical miles. The primary purpose of AIS is for collision avoidance, but Coast Guard officials told us these data can also be used to identify potential IUU fishing. Requirements to broadcast AIS on the high seas depend on the vessel size and the vessel's flag state requirements. AIS data is generally received in real or near-real time and publicly available through a variety of sources, including private companies and a web-based tool developed by the Departments of Transportation and the Navy.
>
> *Vessel Monitoring System (VMS)* is satellite-based technology that automatically broadcasts the location and movement of vessels at a greater range than AIS. However, as with AIS, not all vessels are required to transmit through VMS. VMS requirements vary by flag state and conditions of various international agreements, including Regional Fisheries Management Organizations (RFMO). VMS data are generally treated as proprietary to the flag state of the vessel or RFMO, depending on established flag state and RMFO data rules and procedures.
>
> Additionally, some countries make their vessels' data available through the nonprofit Global Fishing Watch. NOAA is responsible for setting, monitoring, and enforcing VMS use in the U.S.
>
> *Visible Infrared Imaging Radiometer Suite (VIIRS)* is satellite technology that uses highly sensitive optical sensors to see lights at night, enabling visualization of vessels using light to attract catch. According to NOAA officials, VIIRS data is managed by the Earth Observation Group within the Colorado School of Mines.
>
> Source: Information from U.S. Coast Guard, U.S. Department of Transportation, National Oceanic and Atmospheric Administration, Indian Ocean Tuna Commission, and Global Fishing Watch. | GAO-22-104234.

Coast Guard Identifies Potential Incidents of IUU Fishing through Operations at Sea

The Coast Guard identifies potential IUU fishing through at-sea operations, including patrols and boarding and inspections carried out under the authority of RFMOs. The Coast Guard conducts at-sea patrols using its vessels or aircraft. These patrols may be part of missions to address specific types of IUU fishing, such as to identify high seas driftnet fishing. If the Coast Guard observes potential IUU fishing during patrol operations that are not directly related to IUU fishing, it will follow up as appropriate, according to Coast Guard officials. For example, during a mission not focused on IUU fishing, the Coast Guard observed a vessel with a large number of shark fins drying on the deck, which officials told us could be an indicator of illegal shark-finning practices. In response, the Coast Guard provided information on the suspected IUU fishing to NOAA, which then investigated the vessel in cooperation with the Secretariat of the RFMO covering the region of the ocean where the illegal activity was observed.

The Coast Guard may also identify IUU fishing through at-sea boardings and inspections conducted under the authority of RFMO agreements. For example, in 2018 during the Coast Guard's annual fisheries enforcement operation, one Coast Guard patrol vessel conducted boardings of 10 vessels under the authority of the Western and Central Pacific Fisheries Commissions RFMO agreement, which includes a high seas boarding and inspection provision. Of these 10 vessels, the Coast Guard identified six with violations, including not reporting positional data through a vessel monitoring system, as required. When the Coast Guard investigates potential violations through boardings and inspections, it reports them to the RFMO, which alerts the vessel's flag state, according to Coast Guard officials.

Five of the nine RFMO agreements of which the U.S. is a member have high seas boarding and inspection provisions under which member nations may board and inspect vessels flagged to other RFMO member nations to monitor compliance with the RFMO agreement and its conservation measures.[31] From 2016 through 2020, the Coast Guard boarded and inspected 227 fishing vessels on the high seas under RFMO authorities and discovered 90 potential violations of RFMO conservation and management measures, according to information from Coast Guard officials. Coast Guard officials said they had suspected some of these vessels of IUU fishing, and that they

---

[31] The Western and Central Pacific Fisheries Commission, Commission for the Conservation of Antarctic Marine Living Resources, Northwest Atlantic Fisheries Organization, and the North Pacific Fisheries Commission have adopted high seas boarding and inspection provisions that apply to all vessels operating in fisheries managed under their conventions. The International Commission for the Conservation of Atlantic Tunas has adopted boarding and inspection provisions that apply only to Eastern Atlantic and Mediterranean Bluefin tuna and Mediterranean Swordfish fisheries.

randomly boarded others to project an enforcement presence and deter IUU fishing activity.

Coast Guard officials told us they are working with the Department of State and NOAA to promote the adoption of high seas boarding and inspection measures in all RFMO agreements to which the U.S. is a member. Coast Guard officials specified that, in many cases, changing RFMO agreement provisions requires a full consensus among all members, which makes the process challenging.[32] For example, Coast Guard officials said that in 2021, the South Pacific Regional Fisheries Management Organization proposed adopting high seas boarding and inspection provisions, but the initiatives were unable to reach consensus among member countries.

Interagency Groups and Processes Help Coordinate U.S. Efforts to Combat IUU Fishing at Sea, and the Maritime SAFE Act Working Group Generally Followed Selected Leading Collaboration Practices

An interagency working group established by the Maritime SAFE Act in 2019 coordinates U.S. agencies' efforts government-wide to address IUU fishing. We found that this working group generally followed selected leading collaboration practices. In addition, several other interagency groups and processes help coordinate aspects of U.S. efforts related to combating IUU fishing at sea.

Maritime SAFE Act Working Group on IUU Fishing Coordinates U.S. Efforts Government-wide and Generally Followed Selected Leading Collaboration Practices

The Maritime SAFE Act interagency working group coordinates U.S. efforts to combat IUU fishing government-wide and is tasked with ensuring an integrated federal response to IUU fishing globally. The act outlines the working group's responsibilities, which include assessing areas for increased

---

[32] Some RFMOs, such as the Commission for the Conservation of Antarctic Marine Living Resources, make substantive decisions by consensus, while others have voting procedures for some or all decisions, according to NOAA officials. According to NOAA officials, RFMOs favor a consensus-based decision-making process even if it is not required.

interagency information sharing on matters related to IUU fishing; increasing maritime domain awareness relating to IUU fishing;[33] outlining a strategy to coordinate, increase, and use shiprider agreements between DOD or the Coast Guard and relevant countries; and, through a strategic plan, identifying priority regions and priority flag states to be the focus of the working group's assistance.[34] In addition, the act directs the working group to prepare the following documents:

- annual reports summarizing nonsensitive information about the working group's efforts to investigate, enforce, and prosecute groups and individuals engaging in IUU fishing;
- by December 2021, a 5-year integrated strategic plan on combating IUU fishing and enhancing maritime security, including specific strategies with monitoring benchmarks for addressing IUU fishing in priority regions; and
- not later than 5 years after submission of the 5-year integrated strategic plan, and 5 years after, a report on a number of issues related to IUU fishing.

As of June 2021, the group had met four times, established subworking groups and task groups, developed a work plan, and solicited public comments on that work plan.[35] The four subworking groups are to address the following issues:

---

[33] Maritime domain awareness is defined by National Security Presidential Directive 41 (NSPD-41)/Homeland Security Presidential Directive 13 (HSPD-13), *Maritime Security Policy*, issued by the White House in 2004, as the effective understanding of anything associated with the global maritime domain that could impact the security, safety, economy, or environment of the U.S.

[34] The act defines priority regions as those at high risk for IUU fishing activity or the entry of illegally caught seafood into the markets of the countries in the region and in which countries lack the capacity to fully address such illegal activity. It defines priority flag states as countries the flagged vessels of which actively engage in, knowingly profit from, or are complicit in IUU fishing, and that are willing, but lack the capacity, to monitor or take effective enforcement action against their fleet.

[35] The task groups, which the group's work plan states are intended to be narrower in focus and more limited in duration than the subworking groups, address (1) priority regions and priority flag states (led by the State Department), and (2) development of the 5-year strategic plan.

- *Maritime intelligence coordination.* This subworking group of 13 agencies, led by NMIO, has four responsibilities, according to working group documents and NMIO officials:

1) Support the working group in identifying priority regions and priority flag states (including by collating intelligence and supporting development of a framework for analysis, according to a NMIO official).
2) Coordinate completion of an updated interagency memorandum of understanding between the Secretaries of State, Defense, the Interior, Commerce, Homeland Security, and the Office of Director of National Intelligence, on enforcement of U.S. laws and international agreements on living marine resources of the U.S.
3) Leverage the intelligence community to support targeted law enforcement operations and investigations; analyze and share IUU fishing information; and uncover vessel owners, criminal organizations, and flag states that undermine global fisheries management efforts.
4) Lead the establishment of protocols for information sharing and collaboration on emerging technologies and intelligence to support maritime domain awareness and counter-IUU fishing activities.

- *Gulf of Mexico IUU fishing.* Led by NOAA, this subworking group is to identify federal actions taken and policies established during the 5 years prior to enactment of the Maritime SAFE Act with respect to IUU fishing in the U.S. EEZ in the Gulf of Mexico. The subworking group is also to identify actions that NOAA, the State Department, and the Coast Guard can take, using existing resources, to combat IUU fishing in the U.S. EEZ in the Gulf of Mexico, as well as any additional authorities that could assist each agency in more effectively addressing such IUU fishing. Pursuant to the act, NOAA issued a related report to Congress in 2021.
- *Public-private partnerships.* The work plan tasks this subworking group, led by NOAA, with developing a communications strategy and plan to inform and involve stakeholders in the working group's efforts, identifying existing formal partnerships between agencies and private entities, and engaging with existing partners to identify areas for additional efforts.

- Forced labor. Led by the State Department, NOAA, and Department of Labor, this subworking group was formed in 2021 in response to a recommendation in an interagency report to Congress on human trafficking in the seafood supply chain.[36] The report calls for formation of this subworking group to develop and facilitate an integrated approach across the U.S. government to combat human trafficking within the seafood supply chain and to include that work in the working group's 5-year strategic plan.[37]

We found that the Maritime SAFE Act working group's early actions are generally consistent with selected leading collaboration practices we identified in prior work (see fig. 4).[38] For example, the working group's structure includes a rotating leadership role that is shared by NOAA, the Coast Guard, and the State Department. This is consistent with the leading practice of identifying and sustaining leadership. The working group has also developed a work plan, as noted above, that describes the activities of the group and its subworking groups for fulfilling the responsibilities outlined in the Maritime SAFE Act, which is consistent with the leading practice of developing and updating written guidance and agreements. NOAA officials said that the work plan is considered a living document, but that the working group had not yet determined its process for formally updating the work plan over time.

---

[36] NOAA and the State Department, Report to Congress, Human Trafficking in the Seafood Supply Chain, Section 3563 of the National Defense Authorization Act for Fiscal Year 2020 (P.L. 116-92), accessed June 29, 2021, https://media.fisheries.noaa.gov/2020-12/DOSNOA AReport_HumanTrafficking.pdf?null. Among other things, the report listed countries and territories with fisheries or related seafood industries most at risk for human trafficking within their seafood supply chains. One of the report's recommendations was formation of a subworking group on human trafficking, including forced labor, under the Maritime SAFE Act working group.

[37] GAO has also reported on forced labor, including its use in the seafood industry. See GAO, Forced Labor: CBP Should Improve Communication to Strengthen Trade Enforcement, GAO-21-259 (Washington, D.C.: Mar. 1, 2021), and GAO, Forced Labor: Better Communication Could Improve Trade Enforcement Efforts Related to Seafood, GAO-20-441 (Washington, D.C.: June 18, 2020).

[38] GAO-12-1022.

Leading collaboration practices	Examples of issues to consider	Examples of actions taken by the working group
Identifying and sustaining leadership	How will leadership be sustained over the long term? If leadership is shared, have roles and responsibilities been clearly identified and agreed upon?	The working group's leadership consists of a rotating role for three agencies, with a chair term of 3 years. The National Marine Fisheries Service assumed the initial chair role, and agencies plan to identify the next chair before the end of the first term.
Including relevant participants	Have all the relevant participants been included? Do they have the ability to commit resources for their agency?	The heads of 15 departments or agencies from across the federal government are to appoint representatives to the working group, which is also to include representatives from five entities to be appointed by the President, as well as representatives from one or more members of the intelligence community to be appointed by the Director of National Intelligence. Most representatives called for by the act have been appointed.[a]
Developing and updating written guidance and agreements	If appropriate, have participating agencies documented their agreement regarding how they will be collaborating?	The written work plan developed by the working group describes the activities of the group and its subworking group for fulfilling the responsibilities outlined in the Maritime SAFE Act.
Clarifying roles and responsibilities	Have participating agencies clarified roles and responsibilities?	For each subworking group and for other ongoing activities, the work plan identifies a lead agency, other agency participants as applicable, and specific tasks.
Defining outcomes and monitoring accountability	Have short- and long-term outcomes been clearly defined? Is there a way to track and monitor their progress?	The working group established interim outcomes and accountability by developing a work plan with time frames for specific tasks. The group plans to determine how to monitor progress as it begins to develop its 5-year strategic plan, due in December 2021.

Source: GAO analysis of our prior work on leading collaboration practices and information from working group members and documents. | GAO-22-104234.

Notes: The Maritime SAFE Act established a collaborative interagency working group on maritime security and illegal, unreported, and unregulated (IUU) fishing in December 2019. Pub. L. No. 116-92, div. C, tit. XXXV, subtit. C., § 3551, 133 Stat. 1997, 2005 (2019) (codified at 16 U.S.C. § 8031). For our prior work on leading collaboration practices, see GAO, Managing for Results: Key Considerations for Implementing Interagency Collaborative Mechanisms, GAO-12-1022 (Washington, D.C.: Sept. 27, 2012).

[a] According to National Oceanic and Atmospheric Administration officials, the Office of Science and Technology Policy has had turnover of its previously appointed working group representative(s) and a new representative has not yet been appointed.

Figure 4. Maritime Security and Fisheries Enforcement (SAFE) Act Working Group's Implementation of Selected Leading Collaboration Practices, as of June 2021.

Under the act, the heads of 15 specified federal entities are to appoint representatives to the working group. The group is also to include representatives from five entities to be appointed by the President, as well as representatives from one or more members of the intelligence community to be appointed by the Director of National Intelligence. See appendix I for a list of the members specified by the act. Almost all of these entities have appointed representatives to the working group, with the Navy appointing its representative to the group in June 2021.[39] According to DOD officials, the Navy's delay in selecting a representative was due to a number of causes. For instance, DOD officials noted that the Navy's efforts in support of combating IUU fishing involve multiple areas within the Navy, and DOD needed to identify the most appropriate organization to represent the Navy in the working group. DOD officials also said that assisting law enforcement and partner nations in efforts that support combating IUU fishing is not part of the Navy's primary mission.

Our past work found that if collaborative efforts do not consider the input of all relevant stakeholders, important opportunities for achieving outcomes may be missed.[40] The Maritime SAFE Act includes provisions relevant to the Navy, such as calling for (1) the working group to develop a strategy to determine how military assets and intelligence can contribute to enforcement strategies to combat IUU fishing, and (2) agencies to assess opportunities to create partnerships similar to the Oceania Maritime Security Initiative and the Africa Maritime Law Enforcement Partnership in other priority regions.[41] Further, NOAA officials also emphasized the importance of including the Navy as the working group begins efforts to develop the 5-year strategic plan required by the act. Participation in the working group by the Navy's newly appointed member will better position the group to enhance and sustain its collaborative efforts to address IUU fishing worldwide.

Other Interagency Groups and Processes Help Coordinate Some U.S. Efforts Related to IUU Fishing

Other interagency groups and processes help coordinate some broader U.S. efforts that relate to combating IUU fishing at sea. Specifically, agency

---

[39] The Office of Science and Technology Policy has had turnover of its previously appointed working group representative(s); a new representative has not yet been appointed, according to NOAA officials.

[40] GAO-12-1022 and GAO, *Managing for Results: GAO's Work Related to the Interim Crosscutting Priority Goals under the GPRA Modernization Act,* GAO-12-620R (Washington, D.C.: May 31, 2012).

[41] The Oceania Maritime Security Initiative provides that if a Navy vessel is traveling to a priority area in which the Coast Guard needs to conduct counter-IUU fishing work, the Navy may allow Coast Guard officials and law enforcement officials from Pacific Island partner nations to embark on Navy vessels.

officials identified groups focused on maritime intelligence or maritime domain awareness, the Civil Applications Committee, the U.S. Southern Command J2 working group on IUU fishing, and the U.S. Maritime Operational Threat Response Plan as all helping to coordinate U.S. efforts.

Maritime Intelligence or Maritime Domain Awareness Groups

NMIO leads several interagency groups focused on maritime intelligence or maritime domain awareness. According to a NMIO official, these groups serve a broader purpose than addressing IUU fishing; however, the official explained that their efforts can focus on IUU fishing and may facilitate federal efforts to combat it, as appropriate. For example:

- NMIO leads the Maritime Intelligence Strategy Board, which a NMIO official said includes all intelligence agencies and DOD's combatant commands.
- NMIO also leads the Maritime Domain Awareness Executive Steering Committee, which coordinates policies, strategies, and initiatives supporting the nation's maritime domain awareness plan, according to NMIO officials. The committee includes officials from the intelligence community and the Departments of Commerce, State, Defense, Homeland Security, and Transportation, according to NMIO officials.
- NMIO supports the State Department-led the International Maritime Domain Awareness Working Group, according to a NMIO official. This group was created by the Executive Steering Committee and first met in January 2021.[42] Among other things, the group intends to develop mechanisms for government-wide coordination and shared awareness on the U.S.'s international maritime domain awareness efforts. It also intends to gather interagency international priorities, objectives, and requirements to inform the development of a government-wide strategy for global maritime domain awareness.

Civil Applications Committee

Chartered in 1975 and led by the U.S. Geological Survey, the interagency Civil Applications Committee coordinates and oversees the federal civil use

---

[42] According to a working group document, the group's members include the Departments of Commerce (including NOAA), Defense, Homeland Security (including the Coast Guard), Justice, State, and Transportation, as well as the intelligence community.

of classified collections of remotely sensed data.[43] These collections include data collected by military and intelligence capabilities.

Approximately 2 years ago, the committee formed an IUU fishing community of interest that began by coordinating remote sensing resources on IUU fishing, according to State Department and U.S. Geological Survey officials. In addition, the committee is leading ongoing work under the Maritime SAFE Act working group to identify planned or potential geospatial remote sensing technologies that can be leveraged to support maritime domain awareness and capabilities for addressing IUU fishing, according to the work plan for the Maritime SAFE Act working group.

U.S. Southern Command J2 Working Group on IUU Fishing

Coast Guard and DOD officials told us that this combatant command formed a working group in September 2020 to discuss IUU fishing issues related to the command's area of responsibility in the waters adjacent to Central and South America and the Caribbean Sea. The purpose of the group is to promote coordination, information sharing, efforts to address information gaps, and discussions among IUU fishing analysts across the intelligence community and U.S. government, according to DOD officials. These officials said that the group includes representatives from U.S. Southern Command (including Special Operations Command South),[44] the Coast Guard, NOAA, NMIO, the State Department, the Office of Naval Intelligence, and Florida International University.[45] DOD officials said that members of the working group routinely share information on international coordination, initiatives, and events and that the group has also worked with international partners.

---

[43] According to a U.S. Geological Survey official, the committee's principal members are the Coast Guard; Departments of Agriculture, Commerce, Health and Human Services, the Interior, and Transportation; Environmental Protection Agency; Federal Emergency Management Agency; National Aeronautics and Space Administration; National Science Foundation; Tennessee Valley Authority; and U.S. Army Corps of Engineers. Associate members are the Defense Intelligence Agency; Departments of Energy, Homeland Security, and State; National Geospatial-Intelligence Agency; National Guard Bureau; and National Reconnaissance Office. Ex officio members are the Office of the Director of National Intelligence, Office of Science and Technology Policy, and National Geospatial Intelligence Committee.

[44] Special Operations Command South includes military members from all four services. It plans and executes special operations in Central and South America and the Caribbean to find and counter threats to U.S. interests and maintain regional stability.

[45] DOD officials said that Florida International University is developing a Security Research Hub to acquire and collate IUU fishing-related data from various open-source or publicly available information resources, including the nonprofit organization Global Fishing Watch and Windward, a commercial maritime intelligence company.

U.S. Maritime Operational Threat Response (MOTR) Plan

First signed in 2006, MOTR is the presidentially approved plan to achieve coordinated, quick, and decisive U.S. government responses to threats against the U.S. and its interests in the maritime domain. The MOTR plan and its protocols guide federal agencies toward consistent, coordinated, and consensus-based responses to maritime threats, according to the plan. Such threats include illegal fishing, as well as acts of terrorism, piracy, drug trafficking, human trafficking, and arms trafficking. MOTR includes 10 federal entities.[46] In 2010, the Secretary of Homeland Security established the Global MOTR Coordination Center to serve as the national MOTR coordinator. Officials explained that the center is funded by and administratively part of the Coast Guard.

Since inception of the MOTR plan, the U.S. government has used it in more than 1,000 maritime events, including migrant interdictions, drug seizures, terrorism, and piracy, according to MOTR documentation. For example, federal agencies recently used the MOTR plan to address an incident of suspected IUU fishing, according to Coast Guard officials. In this instance, the Coast Guard identified a vessel suspected of IUU fishing in the convention area of an RFMO that had not adopted an applicable high seas boarding and inspection provision. Coast Guard officials said they worked with the State Department to use the MOTR plan as a framework to request authority from the vessel's flag state to board and inspect the vessel. While the flag state ultimately denied this request, U.S. officials and the flag state together determined that the suspected IUU fishing was related to unclear vessel registration processes and resolved the issue. State Department officials said that the resolution included contacting the nation understood to be the vessel's next port of call to request that it perform a port inspection upon the vessel's arrival.

The MOTR plan is also used in exercises to plan for addressing potential incidents of IUU fishing. For example, agencies used it in 2020 to conduct an interagency discussion prior to a Coast Guard IUU fishing patrol, according to State Department officials. State Department officials said that the discussion focused on federal authorities and potential scenarios the Coast Guard could face during its patrol, including legal and policy considerations for potential high seas boardings to combat IUU fishing. The officials said that

---

[46] These include the Centers for Disease Control and Prevention; Departments of Defense, Homeland Security, Justice, State, Transportation, and Treasury; NMIO; NOAA; and National Security Council.

approximately 50 military officers and officials from the Coast Guard, the State Department, and NOAA examined scenarios that could occur during the patrol operations. State Department officials explained that by presenting challenging legal and diplomatic scenarios not normally associated with standard law enforcement and fisheries operations, the exercise provided additional information to U.S. agencies that might become involved in responding to IUU fishing.

Conclusion

IUU fishing at sea is an international issue that causes significant negative economic impacts globally. The U.S. is one of many nations working to combat IUU fishing through international cooperation mechanisms and at-sea exercises to build partner nation capacity to enforce their fisheries law. Additionally, the Maritime SAFE Act provides that it is the policy of the U.S. to develop diplomatic, military, law enforcement, economic, and capacity-building tools to counter IUU fishing, and to promote global maritime security through improved capacity and technological assistance to support improved maritime domain awareness. One such capacity-building tool is DOD's African Maritime Law Enforcement Partnership program, which builds African partner nations' capability to enhance maritime security and enforce their maritime laws at sea through real-world combined maritime law enforcement operations. However, due to legislative changes, DOD lacks clarity on whether it has the authority to fully execute one phase of the partnership—specifically, Operation Junction Rain, which arranges combined maritime law enforcement activities with U.S. and African partner nation personnel. DOD officials told us they do not believe they have the authority to conduct the program, and DOD documents indicate it is not clear whether there may be other available authorities to leverage to continue the operation. According to DOD officials, Operation Junction Rain yielded significant positive results in developing partner nations' capacity to strengthen fisheries law enforcement. If DOD determines whether it has the authority to conduct Operation Junction Rain, the department could either resume the program or seek the necessary authority to do so. Resuming the program would strengthen DOD support of African partner nations in developing their ability to enforce fisheries laws and regulations, which in turn would help them work to counter IUU fishing both in their EEZs and on the high seas when committed by their flagged vessels.

Recommendation for Executive Action

The Department of Defense should determine whether it has the authority to continue to conduct Operation Junction Rain and, if it determines it does not, seek the authority to do so (Recommendation 1).

Agency Comments and Our Evaluation

We provided a draft of this report to the Departments of Commerce, Defense, Homeland Security, and State, and the Office of the Director of National Intelligence for review and comment. We received written comments from the Department of Defense, which are reproduced in appendix II. In addition, NOAA, within the Department of Commerce, the Department of Defense, the Department of Homeland Security, and the Office of the Director of National Intelligence provided technical comments, which we incorporated as appropriate. State Department did not provide comments.

DOD partially concurred with our recommendation. Specifically, DOD stated that it did not believe it should seek any additional authority specifically to conduct law enforcement operations, including enforcement of fishery laws and regulations of African countries. Our recommendation specifies that DOD should determine whether it has the authority to continue to conduct Operation Junction Rain—an operation it previously conducted—and, if it determines it does not, seek the authority to do so. The African Maritime Law Enforcement Partnership program, under which Operation Junction Rain was established, directly supports AFRICOM's efforts to counter IUU fishing, among other things, and DOD officials told us this program yielded significant positive results in the past. We believe that, if DOD determines whether it has the authority to conduct Operation Junction Rain, the Department could either resume the program or seek the requisite authority to do so. Officials told us that resuming the program would strengthen DOD support of African partner nations in developing their ability to enforce fisheries laws and regulations, which in turn would help them work to counter IUU fishing both in their EEZs and on the high seas when committed by their flagged vessels.

As agreed with your offices, unless you publicly announce the contents of this report earlier, we plan no further distribution until 30 days from the report date. At that time, we will send copies to the appropriate congressional committees; the secretaries of Commerce, Defense, Homeland Security, and

State; and the Director of the Office of the Director of National Intelligence. In addition, the report is available at no charge on the GAO website at http://www.gao.gov.

If you or your staff have any questions about this report, please contact me at (202) 512-3841 or JohnsonCD1@gao.gov. Contact points for our Offices of Congressional Relations and Public Affairs may be found on the last page of this report. GAO staff who made key contributions to this report are listed in appendix II.

Cardell D. Johnson
Director, Natural Resources and Environment

Appendix I: Members of the Maritime SAFE Act Working Group on IUU Fishing

The Maritime Security and Fisheries Enforcement (SAFE) Act established a collaborative interagency working group on maritime security and illegal, unreported, or unregulated (IUU) fishing, and specified the membership of this working group.[47] The act provides that there is to be one chair of the working group, which is to rotate between the Commandant of the Coast Guard, the Secretary of State, and the Administrator of the National Oceanic and Atmospheric Administration, on a 3-year term. The act further provides that there are to be two deputy chairs, from a different department than that of the chair, to be appointed from the Coast Guard, the Department of State, and the National Oceanic and Atmospheric Administration. The working group is also to include members from the following 12 federal agencies, to be appointed by their respective agency heads:

---

[47] Pub. L. No. 116-92, div. C, tit. XXXV, subtit. C, § 3551, 133 Stat. 1997, 2005 (2019) (codified at 16 U.S.C. § 8031).

- the Departments of Agriculture, Defense, Justice, Labor, and the Treasury;
- the Federal Trade Commission;
- the Food and Drug Administration;
- the U.S. Navy;
- the U.S. Agency for International Development;
- U.S. Customs and Border Protection;
- the U.S. Fish and Wildlife Service; and
- U.S. Immigration and Customs Enforcement.

The working group is also to include one or more members from the intelligence community,[48] to be appointed by the Director of National Intelligence. This member currently consists of a representative from the National Maritime Intelligence-Integration Office. Finally, the working group is also to consist of representatives, to be appointed by the President, from the following five entities:

- the Council on Environmental Quality;
- the National Security Council; and
- the Offices of Management and Budget, Science and Technology Policy, and the United States Trade Representative.

---

[48] 2 The act uses the definition of intelligence community from section 3 of the National Security Act of 1947 (50 U.S.C. § 3003), which defines such community as consisting of a number of specified agencies and entities.

Appendix II: Comments from the Department of Defense

UNCLASSIFIED

OFFICE OF THE ASSISTANT SECRETARY OF DEFENSE
2400 DEFENSE PENTAGON
WASHINGTON, D.C. 20301-2400

INTERNATIONAL
SECURITY AFFAIRS

Mr. Cardell D. Johnson
U.S. Government Accountability Office
441 G Street, NW
Washington, DC 20548

Dear Mr. Johnson:

This is the Department of Defense (DoD) response to the Government Accountability Office (GAO) Draft Report, GAO 21-104234, "Combatting Illegal Fishing: Clear Authority Could Enhance U.S. Efforts to Partner with Other Nations at Sea," dated August 17, 2021.

The DoD response to the report's recommendation is enclosed. My point of contact is Mr. James Furlo, 703-571-9423, james.a.furlo.civ@mail.mil.

Sincerely,

Deputy Assistant Secretary of Defense for
African Affairs

Enclosure:
As stated

UNCLASSIFIED

UNCLASSIFIED   Enclosure 1

GAO DRAFT REPORT DATED AUGUST 17, 2021
GAO-21-104234

COMBATTING ILLEGAL FISHING: Clear Authorities Could Enhance U.S. Efforts to Partner with Other Nations at Sea

DEPARTMENT OF DEFENSE COMMENTS
TO THE GAO REPORT RECOMMENDATION

GAO RECOMMENDATION: The Department of Defense should determine whether it has the authority to continue to conduct Operation Junction Rain and, if it determines it does not, seek the authority to do so.

DoD RESPONSE: DoD partially concurs. The Department of Defense agrees that it should only undertake operations or activities for which it has sufficient and appropriate legal authority. DoD does not agree with the recommendation that the Department seek additional authority specifically to conduct law enforcement operations, including enforcement of fishery laws and regulations of African countries. As the draft report details, the Department of Homeland Security, through the U.S. Coast Guard, the Department of Commerce, through the National Oceanic and Atmospheric Administration, and the Department of State have important roles in countering illegal, unreported, and unregulated fishing. Multiple organizations within DoD are currently addressing the question of maritime security from both a policy and authorities perspective consistent with DoD's role.

UNCLASSIFIED
1

Appendix III: GAO Contact and Staff Acknowledgments

GAO Contact

Cardell D. Johnson, (202) 512-3841 or johnsoncd1@gao.gov.

Staff Acknowledgments

In addition to the contact named above, Anne-Marie Fennell (Director), Elizabeth Erdmann (Assistant Director), Emily Norman (Analyst in Charge), Krista Breen Anderson, David Dornisch, Will Horowitz, Patricia Moye, Cynthia Norris, Courtney Tepera, Mick Ray, Sara Sullivan, Sarah Veale, Christina Werth, and Sara Younes made key contributions to this report.

GAO's Mission

The Government Accountability Office, the audit, evaluation, and investigative arm of Congress, exists to support Congress in meeting its constitutional responsibilities and to help improve the performance and accountability of the federal government for the American people. GAO examines the use of public funds; evaluates federal programs and policies; and provides analyses, recommendations, and other assistance to help Congress make informed oversight, policy, and funding decisions. GAO's commitment to good government is reflected in its core values of accountability, integrity, and reliability.

Connect with GAO

Connect with GAO on Facebook, Flickr, Twitter, and YouTube. Subscribe to our RSS Feeds or Email Updates. Listen to our Podcasts. Visit GAO on the web at https://www.gao.gov.

To Report Fraud, Waste, and Abuse in Federal Programs

Contact FraudNet:
    Website: https://www.gao.gov/about/what-gao-does/fraudnet Automated answering system: (800) 424-5454 or (202) 512-7700.

Congressional Relations

A. Nicole Clowers, Managing Director, ClowersA@gao.gov, (202) 512-4400, U.S. Government Accountability Office, 441 G Street NW, Room 7125, Washington, DC 20548.

Public Affairs

Chuck Young, Managing Director, youngc1@gao.gov, (202) 512-4800.
    U.S. Government Accountability Office, 441 G Street NW, Room 7149 Washington, DC 20548.

Strategic Planning and External Liaison

Stephen J. Sanford, Managing Director, spel@gao.gov, (202) 512-4707.
    U.S. Government Accountability Office, 441 G Street NW, Room 7814, Washington, DC 20548.

Chapter 4

Combating Illegal Fishing: Better Information Sharing Could Enhance U.S. Efforts to Target Seafood Imports for Investigation*

United States Government Accountability Office

Abbreviations

CBP	Customs and Border Protection
CTAC	Commercial Targeting and Analysis Center
IUU	Illegal, unreported, and unregulated
NOAA	National Oceanic and Atmospheric Administration
NMFS	National Marine Fisheries Service

Why GAO Did This Study

Many illicit activities, such as using prohibited fishing gear, constitute IUU fishing. Such fishing undermines the economic and environmental sustainability of the fishing industry. The illicit nature of IUU fishing means that the size of the problem can be estimated only roughly. However, the U.S. International Trade Commission estimated that about 11 percent of the value of the nation's approximately $22 billion in seafood imports in 2019 were derived from IUU fishing.

---

* This is an edited, reformatted and augmented version of the United States Government Accountability Offic Report to Congressional Requesters, Publication No. GAO-23-105643, dated May 2023.

In: Ongoing Efforts to Combat Illegal, Unreported …
Editor: Gordon B. Maddox
ISBN: 979-8-89530-858-5
© 2026 Nova Science Publishers, Inc.

GAO was asked to review federal efforts to combat imports of seafood caught through IUU fishing. This report (1) describes NMFS and CBP efforts to combat such imports; and (2) examines the mechanisms these agencies use to share information on such imports, and related challenges they have identified.

GAO reviewed documents on NMFS and CBP efforts to combat imports of seafood caught through IUU fishing, as well as other relevant agency documents. GAO also interviewed officials from CBP and NMFS at the headquarters and port levels about these efforts and how officials share relevant information on seafood imports.

What GAO Recommends

GAO recommends that CBP work with NMFS to ensure that NMFS has timely access to information that it needs to combat imports of seafood caught through IUU fishing. The Department of Homeland Security concurred with our recommendation.

What GAO Found

The National Marine Fisheries Service (NMFS) and the Department of Homeland Security's U.S. Customs and Border Protection (CBP) work to combat imports of seafood caught through illegal, unreported, and unregulated (IUU) fishing, which comprises many illicit activities (see fig.). For example, NMFS administers four trade monitoring programs that, by regulation, require documentation for imports of specific species. In addition, both agencies manage efforts to identify or "target" seafood imports potentially caught through IUU fishing so that such imports can be investigated or held for further inspection. Targeting efforts can include monitoring incoming seafood imports that fit a pattern of concern, such as importers with past trade violations.

CBP and NMFS share information with each other through several mechanisms, including a data analysis tool and a CBP interagency coordination center.

Common Types of Illegal, Unreported, and Unregulated Fishing

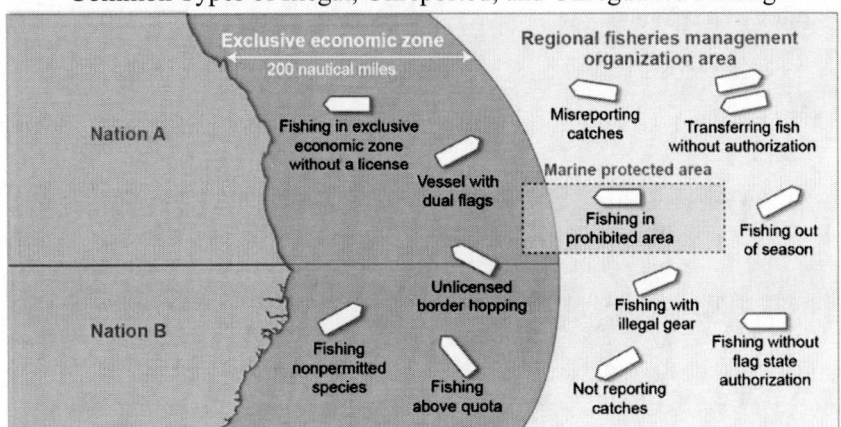

Source: GAO analysis of agency information. | GAO-23-105643.

However, NMFS officials report difficulties obtaining timely information from CBP. In particular, NMFS officials told GAO that having to request information through CBP's interagency coordination center limits their ability to get the timely information they need. NMFS officials told us that if they want to inspect an incoming shipment, they need sufficient advance notice to coordinate with CBP. In addition, they said that receiving the information they need from CBP's interagency coordination center, in some cases, could take as long as a week.

CBP officials told us that coordinating information requests through its coordination center is important to ensure that such requests do not jeopardize or duplicate ongoing CBP operations, among other reasons. Until CBP works with NMFS to ensure timely access to needed information, both NMFS and CBP may miss opportunities to combat imports of seafood caught through IUU fishing.

May 19, 2023

The Honorable Dan Sullivan
Ranking Member
Subcommittee on Oceans, Fisheries, Climate Change,
and Manufacturing
Committee on Commerce, Science, and Transportation
United States Senate

The Honorable John Thune
nited States Senate

The Honorable Roger F. Wicker
United States Senate

Illegal, unreported, and unregulated (IUU) fishing is a global problem that may undermine the economic and environmental sustainability of fisheries, jeopardize food and economic security, and benefit transnational crime. IUU fishing encompasses many illicit activities, ranging from underreporting the number and types of fish caught to using prohibited fishing gear, such as unauthorized driftnets.[1] The U.S. is one of the largest seafood import markets globally and is the second-largest consumer of seafood.[2] The National Marine Fisheries Service (NMFS) estimates that the U.S. imports from 70 to 85 percent of the seafood consumed here[3]—in 2021, about 7.3 billion pounds of imported seafood valued at nearly $28.5 billion, according to our analysis of foreign trade data.[4]

Although the illicit nature of IUU fishing means that the size of the problem and its negative consequences can be estimated only roughly, the U.S. International Trade Commission estimated that about 11 percent of the value

---

[1] Driftnets are large nets designed to drift with the current and entangle fish in the nets' webbing.

[2] U.S. International Trade Commission, *Seafood Obtained via Illegal, Unreported, and Unregulated Fishing: U.S. Imports and Economic Impact on U.S. Commercial Fisheries*, Publication Number 5168 (Washington, D.C.: Feb. 2021).

[3] NMFS is within the Department of Commerce's National Oceanic and Atmospheric Administration (NOAA). The imported seafood in the estimate may include a substantial portion of domestic catch exported for processing and then imported to the U.S. in a processed form—for example, lobster harvested in the Northeast U.S. and processed in Canada; salmon and other finfish, such as pollock and cod captured off the coast of Alaska and processed in China; and tuna caught by U.S. vessels in the Pacific Ocean and processed in Asia and South America. See U.S. Department of Commerce, National Marine Fisheries Service, *Fisheries of the United States, 2019*, NOAA Current Fishery Statistics No. 2019 (Silver Spring, MD: 2021); and United States International Trade Commission, *Seafood Obtained via Illegal, Unreported, and Unregulated Fishing*.

[4] National Marine Fisheries Service, Office of Science and Technology, Foreign Trade Query, available at: http://www.fisheries.noaa.gov/foss, accessed October 5, 2022. At the time of our analysis, 2021 was the most recent full year for which data were available.

According to NMFS, the agency purchases foreign trade data from the Foreign Trade Division of the U.S. Census Bureau. The bureau receives the data from U.S. Customs and Border Protection, which receives the data from importers and exporters. Our analysis of these data indicates that 94 percent of the weight of fishery products imported in 2021—and 98 percent of their dollar value—were edible products. In this report, the term "seafood" may include the small amount of inedible imported fishery products, such as fish meal.

of the nation's $22 billion in seafood imports in 2019 was derived from IUU fishing.[5] Various factors can make it difficult to identify and intercept such imports, including the volume and variety of species imported and the complex supply chains of imported seafood.[6]

In 2021, we reported on federal efforts to combat IUU fishing at sea, including how the U.S. works with other nations, identifies potential incidents of IUU fishing, and coordinates its related interagency efforts.[7] Subsequently, you asked us to review federal efforts at U.S. ports to combat imports of seafood caught through IUU fishing. NMFS leads federal efforts to combat IUU fishing and works in coordination with U.S. Customs and Border Protection (CBP), within the Department of Homeland Security, to address U.S. imports of seafood potentially caught through IUU fishing. This report (1) describes NMFS and CBP efforts to combat such imports; and (2) exami to share information with each other on such imports, and related challenges they have identified.

To describe NMFS and CBP efforts to combat imports of seafood caught through IUU fishing, we reviewed documents and interviewed officials from each agency to determine what programs are in place and how they operate. We reviewed NMFS program reports and forms for imports, guidelines on information that importers must provide, and the agencies' written responses to our questions. We interviewed relevant agency officials about each agency's roles and responsibilities with respect to seafood imports. For example, depending on the agency, we discussed how the agency screens seafood shipments before entry to commerce, inspects random or selected shipments in ports, and identifies—through audits or other investigations—imports potentially caught through IUU fishing once those shipments have entered U.S. commerce.

To examine the mechanisms that NMFS and CBP use to share information with each other on seafood imports, and related challenges that they have

---

[5] U.S. International Trade Commission, *Seafood Obtained via Illegal, Unreported, and Unregulated Fishing*. This report was prepared in response to a congressional request in 2021, and the U.S. International Trade Commission has not published any subsequent reports focused on seafood obtained via IUU fishing.
[6] For example, catches from several smaller boats may be combined at sea onto a bigger vessel before transport to shore for processing. This practice may combine legal and illegal catches before they are landed. In addition, multiple seafood harvests may be processed together into seafood products, which could also combine legal and illegal catches. Seafood products may then be combined and divided multiple times as they change hands at wholesale facilities before further distribution.
[7] GAO, *Combating Illegal Fishing: Clear Authority Could Enhance U.S. Efforts to Partner with Other Nations at Sea*, GAO-22-104234 (Washington, D.C.: Nov. 5, 2021).

identified, we reviewed documents and interviewed officials from each agency. We reviewed interagency agreements, documentation on relevant CBP data systems and analysis tools, related correspondence between agencies, and NMFS' and CBP's written responses to our questions about their information-sharing mechanisms. In our interviews with relevant agency officials, we discussed mechanisms that they use to share information to combat imports of seafood caught through IUU fishing, challenges that they face in sharing such information, and potential opportunities that they identified to improve coordination. We evaluated one of the information-sharing mechanisms that agency officials identified against a relevant interagency memorandum of understanding.[8]

We also interviewed agency officials from two selected ports to (1) describe how the agencies coordinate to combat imports of seafood caught through IUU fishing; and (2) identify examples of effective coordination, coordination challenges that the agencies face, and potential opportunities for improving coordination, as identified by these agencies. To select these ports, we first selected two customs districts, with one chosen to represent a customs district with a high volume of seafood imports and the second chosen to represent a medium-volume customs district.[9] Because the customs districts include multiple ports, we then selected the ports with the same names as the customs districts. We also incorporated geographic diversity by selecting districts/ports on both the East and West Coasts of the U.S. The selection of ports does not provide a generalizable sample of U.S. ports.

We conducted this performance audit from January 2022 to May 2023 in accordance with generally accepted government auditing standards.

Those standards require that we plan and perform the audit to obtain sufficient, appropriate evidence to provide a reasonable basis for our findings and conclusions based on our audit objectives. We believe that the evidence obtained provides a reasonable basis for our findings and conclusions based on our audit objectives.

---

[8] See Customs and Border Protection et al., Addendum to Import Safety Commercial Targeting and Analysis Center Memorandum of Understanding Among the Import Safety Commercial Targeting and Analysis Center, Partner Government Agencies (2017).

[9] According to CBP officials, the term "customs district" is no longer used, and ports are now organized by CBP field offices; however, at the time of our selecting ports, the data on seafood imports were organized by customs districts.

Background

Definition of IUU Fishing

The importation of fish harvested through IUU fishing is prohibited under several federal laws, including the Magnuson-Stevens Fishery Conservation and Management Act and the Lacey Act.[10] IUU fishing is a broad term that generally includes activities that violate national law or international fishing regulations or agreements.[11] The Department of Commerce's National Oceanic and Atmospheric Administration (NOAA) generally describes each aspect of IUU fishing as follows:[12]

- Illegal fishing refers to fishing activities conducted in contravention of applicable laws and regulations, including those adopted at the regional and international level.
- Unreported fishing refers to fishing activities that are not reported or are misreported to relevant authorities in contravention of national laws and regulations or reporting procedures of a relevant regional fisheries management organization.[13]

---

[10] The Magnuson-Stevens Fishery Conservation and Management Act makes it unlawful to import, export, transport, sell, receive, acquire, or purchase in interstate or foreign commerce any fish taken, possessed, transported, or sold in violation of any foreign law or regulation or any treaty in contravention of any binding conservation measure adopted by an international agreement or organization to which the U.S. is a party. 16 U.S.C. § 1857(1)(Q). The Lacey Act prohibits the import, export, transport, sale, receipt, acquisition, or purchase in interstate or foreign commerce of any fish taken, possessed, transported, or sold in violation of any law, treaty, or regulation of the U.S., or in violation of any foreign law. 16 U.S.C. § 3372(a). Additionally, several statutes implementing the various regional fisheries management organization conventions to which the U.S. is a party provide enforcement tools to address imports of IUU fish. *See, e.g.*, Tuna Conventions Act of 1950, 16 U.S.C. § 957(a); Antarctic Marine Living Resources Convention Act, 16 U.S.C.§§ 2435, 2437, 2439; Northwest Atlantic Fisheries Convention Act, 16 U.S.C. § 5606.

[11] Applicable requirements for fishing vessels at sea and, thus, the kinds of fishing that are permissible, vary depending on the maritime zone. A nation's territorial waters generally extend from a nation's coastline up to 12 nautical miles away. Beyond and adjacent to the territorial sea, coastal nations generally have an exclusive economic zone up to 200 nautical miles from their coastlines. Beyond exclusive economic zones, the ocean is generally defined as "high seas" and is considered international waters.

[12] NOAA's description of IUU fishing is based on definitions from the *International Plan of Action to Prevent, Deter, and Eliminate Illegal, Unreported and Unregulated Fishing*, adopted at the 24th Session of the Committee on Fisheries in Rome on March 2, 2001. As noted previously, NMFS is a component of NOAA.

[13] Regional fisheries management organizations are treaty-based international bodies comprising nations that share an interest in managing and conserving fisheries in specific regions of the

- Unregulated fishing occurs in geographic areas or for specific species of fish for which there are no applicable conservation or management measures and when fishing activities are conducted in a manner inconsistent with a nation's responsibilities for the conservation of living marine resources under international law. Fishing activities are also unregulated when occurring in an area managed by a regional fisheries management organization and conducted by vessels without nationality, or by those flying a flag of a nation or fishing entity that is not party to the organization, in a manner inconsistent with the conservation measures of that organization.

IUU fishing encompasses many illicit activities that can occur both within a nation's exclusive economic zone as well as on the high seas. For example, these could include fishing without an appropriate license, above a nationally established quota, out of season, or in a prohibited area. Figure 1 below illustrates common types of IUU fishing.

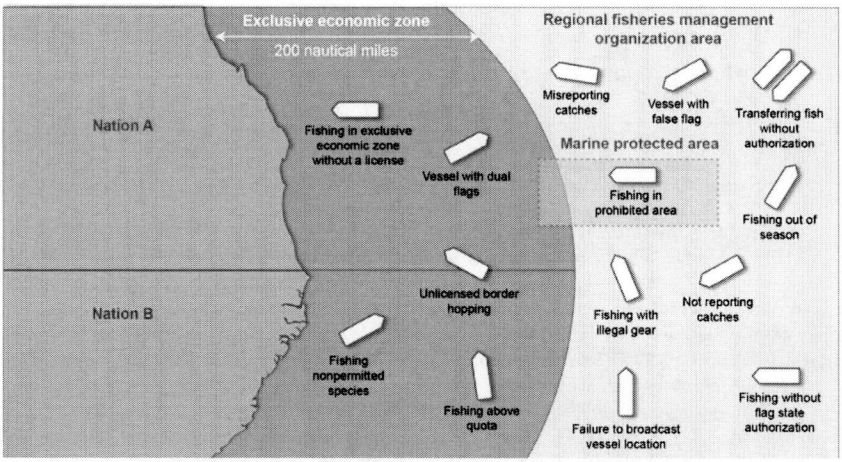

Source: GAO analysis of agency information. | GAO-23-105643.

Figure 1. Common Types of Illegal, Unreported, and Unregulated Fishing.

---

high seas. The U.S. belongs to nine such organizations where the U.S. is a coastal nation or has a fishing interest, according to State Department officials.

U.S. Efforts to Combat IUU Fishing before Seafood Reaches Ports

The U.S. combats IUU fishing at different stages of the seafood supply chain, including efforts designed to detect and combat IUU fishing where it occurs. For example, as we reported in 2021, the U.S. conducts at-sea operations focused on strengthening other nations' capacity to manage their own fisheries and fleets. The U.S. also works with other nations through various multilateral agreements.[14] One such agreement is the Port State Measures Agreement—the first binding international agreement to specifically target IUU fishing, according to the United Nations Food and Agriculture Organization.[15] Among other provisions, the agreement seeks to block fishery products derived from IUU fishing from reaching markets by denying port access to vessels known to have engaged in such fishing.

U.S. Seafood Imports

Globally, the U.S. seafood import market is one of the largest and most diverse in terms of products, according to estimates from the United States International Trade Commission.[16] In 2020, shrimp, salmon, and tuna were the highest valued U.S. seafood imports, according to NMFS.[17] Among these, shrimp was the most valuable, accounting for 27 percent of the value of total edible U.S. seafood imports.[18] NMFS officials stated that the majority of U.S. seafood imports are fresh or frozen products; accordingly, they must be kept refrigerated or frozen during shipping and at port, including any time spent awaiting inspection.

---

[14] GAO-22-104234.

[15] The United Nations Food and Agriculture Organization approved the Agreement on Port State Measures to Prevent, Deter and Eliminate Illegal, Unreported and Unregulated Fishing in 2009, and the agreement entered into force on June 5, 2016. Seventy-four nations, including the U.S., have become parties to the agreement.

[16] United States International Trade Commission, Seafood Obtained via Illegal, Unreported, and Unregulated Fishing.

[17] U.S. Department of Commerce, National Marine Fisheries Service, *Fisheries of the United States, 2020*, NOAA Current Fishery Statistics No. 2020 (Silver Spring, MD: 2022).

[18] U.S. Department of Commerce, National Marine Fisheries Service, *Fisheries of the United States, 2020*.

According to our analysis of foreign trade data, 51 percent of U.S. seafood imports by weight in 2021 were from Asia. As shown in figure 2, China and India together provided 23 percent of the total imports by weight.[19]

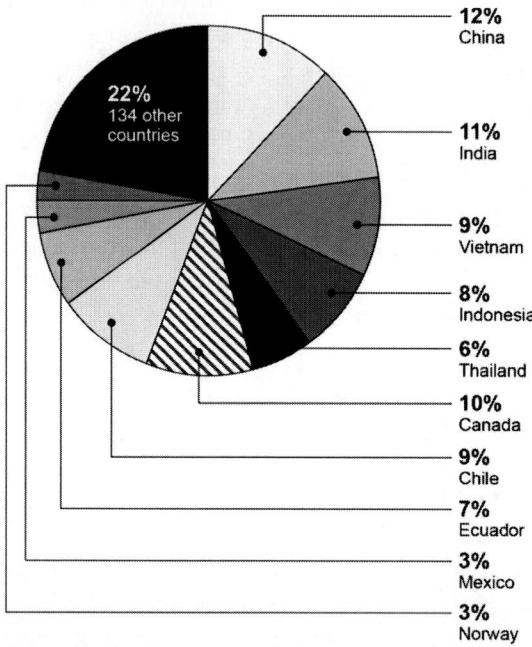

Source: GAO analysis of foreigh trade data from the National Marine Fisheries Service (NMFS) Office of Science and Technology, Foreign Trade Query, available at: http://www. Fisheries.noaa.gov/foss, accessed October 5, 2022. | GAO-23105643.

Note: Countries shown are the top 10 countries from which the U.S. imported seafood in 2021, ranked by the weight of the seafood. In addition to edible seafood, the data include inedible imported fishery products, such as fish meal; such products account for 6 percent of the weight of seafood imported by the U.S. in 2021. According to NMFS, the agency purchases foreign trade data from the Foreign Trade Division of the U.S. Census Bureau. The bureau receives the data from U.S. Customs and Border Protection, which receives the data from importers and exporters.

Figure 2. Origins of Seafood Imported by the U.S. in 2021.

---

[19] National Marine Fisheries Service Office of Science and Technology, Foreign Trade Query, available at: http://www.fisheries.noaa.gov/foss, accessed October 5, 2022. At the time of our analysis, 2021 was the most recent full year for which data were available.

Imported goods enter the U.S. by sea, air, and land at more than 300 ports. NMFS officials told us that most seafood imported into the U.S. arrives at seaports in container vessels, but some high-value seafood arrives by air. The majority of seafood imports enter the U.S. through a small number of customs districts, each of which can comprise multiple ports. According to our analysis of foreign trade data, 10 customs districts received 84 percent of the weight of U.S. seafood imports in 2021, with the New York and Los Angeles districts together accounting for 40 percent of seafood imports.[20]

Agency Roles and Responsibilities in the Seafood Import Process

NOAA's mission includes conserving coastal and marine ecosystems and resources. NOAA addresses the importation of illegally harvested fish and fish products, which, as noted previously, is prohibited under several federal laws, including the Magnuson-Stevens Fishery Conservation and Management Act and the Lacey Act.[21] Within NOAA, NMFS leads federal efforts to combat IUU fishing and has responsibilities for certain seafood imports. NMFS' mission includes providing for productive and sustainable fisheries, safe sources of seafood, recovery and conservation of protected resources, and healthy ecosystems. Combating IUU fishing is one of NMFS' top priorities, according to its strategic plan.[22]

CBP's mission priorities include facilitating lawful international trade at the ports-of-entry for imports, including seafood, and protecting revenue. Accordingly, the agency's responsibilities include facilitating and enforcing the import process and collecting the duties, taxes, and fees assessed on products, including seafood. Additionally, CBP officials told us that the agency conducts enforcement efforts for trade violations, including assessing civil penalties and supporting criminal prosecutions. CBP collects import data and documentation from importers—including a description of the product, manufacturer information, and the country of origin—through its Automated Commercial Environment system. This system and other sources provide data that CBP and its partner agencies, including NMFS, can use to identify cargo

---

[20] National Marine Fisheries Service Office of Science and Technology, Foreign Trade Query, available at: http://www.fisheries.noaa.gov/foss, accessed October 5, 2022. At the time of our analysis, 2021 was the most recent full year for which data were available.

[21] *See* 16 U.S.C. §§ 1857(1)(Q), 3372(a).

[22] U.S. Department of Commerce, National Marine Fisheries Service, *NOAA Fisheries Strategic Plan 2022-2025* (Silver Spring, MD: Dec. 2022).

for possible inspection by port officials.[23] CBP requires certain entry data and documents for all imported goods. Other agencies with regulatory responsibilities for imports may also specify data that importers or their customs brokers need to submit.[24]

Other agencies also have certain roles and responsibilities related to seafood imports, but officials from these agencies told us that their programs do not explicitly focus on combating IUU fishing or imports of seafood caught through IUU fishing. For example:

- The U.S. Fish and Wildlife Service monitors wildlife trade and works to prevent the illegal import or export of species regulated under international agreement and U.S. wildlife laws and regulations. Virtually all wildlife imports and exports must be declared to the Fish and Wildlife Service and cleared by the agency's wildlife officers.
- The Food and Drug Administration is responsible for ensuring that seafood products imported into the U.S. for consumption are safe, sanitary, wholesome, and honestly labeled.

NMFS and CBP Implement Trade Monitoring Programs, Targeting Efforts, and Inspections to Combat Imports of Seafood Caught through IUU Fishing

To combat imports of seafood caught through IUU fishing, NMFS and CBP implement trade monitoring programs, targeting efforts, and inspections. NMFS administers four trade monitoring programs that may help deter and identify imports of seafood caught through IUU fishing by requiring importers to submit specific data and documentation on seafood shipments arriving at

---

[23] According to CBP officials, under CBP policy, agencies that require access to Automated Commercial Environment data to carry out their responsibilities must enter into a memorandum of understanding with CBP. According to CBP officials, the memorandum is to specify the data that the agency is authorized to receive and the ways that the data will be transmitted to the agency. Authorized staff of these agencies receive access to the specified data.

[24] According to CBP, customs brokers assist importers and exporters in meeting federal requirements governing imports and exports. The brokers, who are licensed and regulated by CBP, submit information and appropriate payments to CBP on behalf of their clients and charge a fee for this service, according to CBP.

U.S. ports of entry.[25] CBP and NMFS both have efforts to target imports of seafood potentially caught through IUU fishing—that is, to evaluate, monitor, and identify shipments that fit a pattern of interest or concern, such as importers with past trade violations, according to NMFS officials. Additionally, CBP and NMFS officials at ports conduct some physical inspections that may identify or deter imports of seafood caught through IUU fishing.

NMFS Trade Monitoring Programs May Help Deter and Identify Seafood Imports Caught through IUU Fishing

NMFS has four trade monitoring programs that require documentation and include reviews that NMFS officials told us help deter and identify imports of seafood caught through IUU fishing. These are the Antarctic Marine Living Resource Program, the Highly Migratory Species International Trade Program, the Seafood Import Monitoring Program, and the Tuna Tracking and Verification Program. Each program has certain requirements, set forth in NOAA regulations, for filing certain information about specific marine species, for example, those more vulnerable to IUU fishing or that may be caught through fishing practices that can harm other species. Some species are subject to the requirements of multiple programs.[26] See figure 3 for a summary of the species covered and the documentation reviews that NMFS conducts under each program.

Documentation requirements. Information required under the trade monitoring programs varies by program but may include where the harvest took place, the weight of the seafood being imported, fishing permit or license numbers, the vessel's flag country, harvesting gear type, and dates of harvest.[27] In general, importers or their customs brokers must submit information on

---

[25] NMFS officials told us that one of its four trade monitoring programs—the Tuna Tracking and Verification Program—may also help deter and identify domestic processing of tuna caught through IUU fishing. According to agency officials, the program (1) conducts spot checks by randomly purchasing tuna products labeled as dolphin-safe from retail stores and tracing the tuna's origins and (2) conducts audits at U.S. canneries to determine how the canneries track the dolphin-safe status of products as they move through processing.

[26] *See* 50 C.F.R. pt. 216, subpt. H; pt. 300, subpts. G, M, Q; pt. 635.

[27] NMFS' seafood import and export tool allows users to identify which trade monitoring program or programs that a particular species and product fall under and, therefore, which requirements such species and product are subject to. For the tool, see http://www.fisheries.noaa.gov/seafood-import-export-tool.

incoming shipments into CBP's Automated Commercial Environment system at the time of importation.

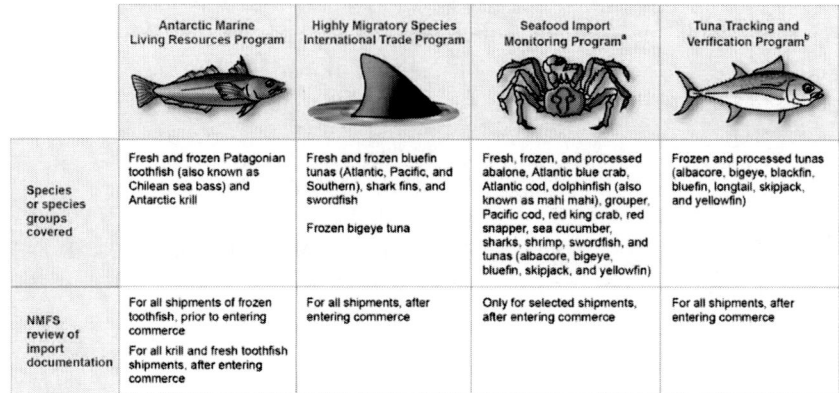

	Antarctic Marine Living Resources Program	Highly Migratory Species International Trade Program	Seafood Import Monitoring Program[a]	Tuna Tracking and Verification Program[b]
Species or species groups covered	Fresh and frozen Patagonian toothfish (also known as Chilean sea bass) and Antarctic krill	Fresh and frozen bluefin tunas (Atlantic, Pacific, and Southern), shark fins, and swordfish Frozen bigeye tuna	Fresh, frozen, and processed abalone, Atlantic blue crab, Atlantic cod, dolphinfish (also known as mahi mahi), grouper, Pacific cod, red king crab, red snapper, sea cucumber, sharks, shrimp, swordfish, and tunas (albacore, bigeye, bluefin, skipjack, and yellowfin)	Frozen and processed tunas (albacore, bigeye, blackfin, bluefin, longtail, skipjack, and yellowfin)
NMFS review of import documentation	For all shipments of frozen toothfish, prior to entering commerce For all krill and fresh toothfish shipments, after entering commerce	For all shipments, after entering commerce	Only for selected shipments, after entering commerce	For all shipments, after entering commerce

Source: GAO analysis of program documents and information from National Marine Fisheries Service (NMFS) officials; GAO (illusitrations). | GAO-23-105643.

[a] On December 28, 2022, NMFS issued a proposed rule to expand the Seafood Import Monitoring Program. 87 Fed. Reg. 79,836 (Dec. 28, 2022). The proposed rule would, among other things, (1) expand the currently listed red snapper and tuna to larger species groups; and (2) add cuttlefish and squid, eels, octopus, queen conch, and Caribbean spiny lobster to the program.

[b] This program is also known as the Dolphin-safe Program.

Figure 3. Species and Documentation Reviews under the National Marine Fisheries Services' Trade Monitoring Programs.

NMFS officials told us that the trade monitoring programs' extensive documentation requirements can deter imports of seafood caught through IUU fishing. According to agency officials, importers or customs brokers knowingly attempting to import seafood caught using IUU fishing would not be able to do so without deliberate misrepresentation or falsification of required documentation. Additionally, an official told us that the program's documentation requirements help ensure that importers have awareness of their own supply chains.

Documentation reviews by NMFS staff. According to agency officials, NMFS reviews documentation for a selection of shipments under the Seafood Import Monitoring Program and for all shipments under the agency's other three trade monitoring programs to ensure compliance with program requirements.[28] NMFS generally reviews shipments after they enter commerce.[29]

NMFS officials said that these reviews are the most efficient method of identifying seafood imports that may have been caught through IUU fishing. Officials told us that these reviews provide an overview of the import process as a whole and that routine and organized document review provides a means to identify seafood potentially caught through IUU fishing among the immense volume of total seafood imports. This overview can help NMFS identify shipments that may be of concern based on missing or inconsistent information, such as vessel permit dates that do not match harvest dates, or other indicators of potential violations. Officials told us that these reviews can inform future investigations and may result in penalties or other consequences for noncompliant importers or brokers.

If NMFS officials identify an issue that is of a more serious nature, or that remains unresolved after outreach, they refer the issue to NOAA's Office of Law Enforcement. Such referrals are relatively uncommon, however, according to NMFS officials. For example, according to a 2021 NMFS report, the majority of documentation reviews under the Seafood Import Monitoring Program do not identify noncompliance; of the 40 percent that did so during the program's first 2 years of implementation, only a small number were significant enough to be referred for enforcement action.[30] NMFS officials said that enforcement actions may include informal outreach or education to support future compliance, written warnings, summary settlements with penalties, referral for civil administrative enforcement, or referral to the Department of Justice for criminal prosecution.

---

[28] In addition, CBP officials said that upon request from NOAA's Office of Law Enforcement, CBP's Regulatory Audit Directorate assists the Seafood Import Monitoring Program with audits of seafood importers.

[29] Frozen toothfish—also known as Chilean sea bass—subject to the Antarctic Marine Living Resources Program are the only seafood imports for which NMFS reviews all documentation prior to the shipments' entry to commerce. NMFS officials said that while pre-entry review is practical for frozen toothfish, which require pre-entry approval, it is impractical for the other programs due in part to their sheer volume of products and more detailed reviews. For example, NMFS officials noted that the Antarctic Marine Living Resources Program covers about 1 percent of seafood products entering the U.S., whereas the Seafood Import Monitoring Program covers about 45 percent of U.S. seafood imports. NMFS officials also cited other factors, including that the number of vessels harvesting Antarctic toothfish is limited, the product is all coming from the same ocean area, and frozen toothfish are shipped by container, so NMFS receives 4 to 6 weeks advance notice of a shipment destined for the U.S.

[30] U.S. Department of Commerce, National Marine Fisheries Service, *Report on the Implementation of the U.S. Seafood Import Monitoring Program* (Silver Spring, MD: 2021).

Officials noted that when they find something concerning, it is not always an indication of IUU fishing or fraud. Instead, it is more typically an issue of the importer unintentionally failing to meet the documentation and reporting requirements or an inconsistency in the information reported.

CBP and NMFS Manage Various Efforts to Target Imports of Seafood Potentially Caught through IUU Fishing

At both the headquarters and port levels, CBP and NMFS work to target imports of seafood potentially caught through IUU fishing. Officials told us that CBP's primary effort for targeting such imports is its Commercial Targeting and Analysis Center (CTAC), an interagency coordination center that combats a broad array of illegal imports beyond seafood.

Comprising 12 partner agencies, including NMFS, CTAC develops and implements procedures that help target seafood imports caught through IUU fishing.

Officials told us that partner agencies bring intelligence to CTAC, which can then facilitate further discussion and analysis and establish import alerts in CBP's data systems to help target shipments that meet specified criteria. Once a target is identified, CTAC coordinates with individual ports to hold and inspect the shipment. For example, CBP officials told us that CBP and NMFS developed alerts for shipments of tuna imported from a specific country due to a suspicion of concealed IUU fishing. This joint effort found that almost all tuna imported from that country violated import requirements, such as correct labeling, under the Seafood Import Monitoring Program. Among other actions in response, CBP took actions to recover nearly $600,000 in underpaid tariffs.

Besides CTAC, CBP officials told us that other CBP entities can contribute to targeting efforts in the following ways:[31]

- *Intelligence Division:* can generate intelligence reports on IUU fishing and provide those to CBP or partner agencies for use in targeting shipments.

---

[31] In addition, an official from the Department of Homeland Security's Immigration and Customs Enforcement—a CTAC partner agency—told us that this agency's Homeland Security and Investigations directorate makes significant contributions to CTAC. Homeland Security and Investigations conducts investigations of IUU fishing to support national and international prosecutions and provide evidence to guide future policy and management of fisheries.

- *Strategic Enforcement Branch:* works with CBP and partner agencies to provide intelligence and analysis of anomalies in supply chains and prior shipments, investigate trade violations, target shipments, and take enforcement actions. The branch also provides related training to CBP port officials.
- *National Threat Analysis Centers (Miami):* can develop initial national-level targeting efforts for seafood imports and create alerts in CBP's systems for targeted shipments.
- *Agricultural and Prepared Products Center for Excellence and Expertise:* can look for patterns in violations that may indicate a systemic problem with an industry or importer, and help other CBP entities develop targeting plans.
- National Targeting Center: develops targeting policy and is responsible for certain data tools that CBP and its partner agencies use to access and analyze data on imports.

NMFS and CBP officials at the port level may also initiate efforts to target imports of seafood potentially caught through IUU fishing. Specifically, NMFS and CBP officials operating at the two ports we selected told us that they can initiate targeting based on local intelligence and work with partner agencies and national-level CBP groups, including CTAC, to further develop or share information. For example, CBP officers we interviewed at one port told us that they generate intelligence that is used locally and shared nationally with CBP through CTAC. Once shared with CTAC, this local intelligence helps inform future targeting efforts at the national level, according to those officers.

CBP and NMFS Port Officials Jointly Inspect Some Seafood Imports

NMFS and CBP jointly conduct physical inspections of some seafood imports upon arrival at U.S. ports to potentially identify seafood caught through IUU fishing. NMFS and CBP officials told us that, in general, NMFS officials in ports and at headquarters identify shipments for inspection based on targeting efforts, although some inspections may be random. Once a shipment is identified for inspection, NMFS officials request that CBP hold the shipment at the port.[32] NMFS and CBP officials then open the shipping containers;

---

[32] According to NMFS officials, NMFS does not have the authority to hold shipments.

visually examine the shipment, and compare it with the entry information supplied by the importer.

In some cases, NMFS' Marine Forensic Laboratory can analyze a sample from a shipment to verify the species and may also be able to determine its geographic origin and whether a product was wild caught or farm raised. For example, in a 2021 case that NMFS ultimately referred to the Department of Justice for criminal prosecution, lab results determined that three shipping containers of shrimp were misrepresented as farm raised when they were actually wild caught, according to NMFS officials.

According to NMFS officials, this misrepresentation was a violation of U.S. import prohibitions under the Department of State's Section 609 program, which is designed to protect sea turtles in foreign shrimp fisheries.[33]

The number of inspections is low relative to the volume of such imports, according to NMFS officials. While NMFS does not track the number of inspections, officials told us that in November 2021, NMFS conducted approximately 14 import inspections—12 at seaports and two through truck inspections at border crossings—and that this represented a typical number of inspections.[34] NMFS officials identified several reasons for the limited number of inspections:

- The benefit of inspecting a particular seafood shipment must be balanced against the risk of impeding the flow of thousands of other unrelated shipments, especially in busy ports.
- NMFS has limited capacity to conduct import inspections; there are 22 NMFS law enforcement officers who conduct these efforts in the more than 300 ports.
- There is no advance schedule for when seafood imports will arrive by land.

---

[33] Under this program, shrimp harvested with commercial fishing technology that may adversely affect sea turtles cannot be imported into the U.S., except that the import ban does not apply if the President and Congress certify that the government of the harvesting nation has adopted certain protective measures. *See* Pub. L. No. 101-162, § 609, 103 Stat. 1037 (1989) (16 U.S.C. § 1537 note).

[34] NMFS officials told us that a number of factors complicate their ability to quantify the number of shipments inspected, including inspections of cargo containers that may involve multiple shipments. Additionally, officials told us that while the agency did not currently track the number of inspections, an upcoming update to NMFS' case tracking system may better facilitate tracking inspections in the future.

Officials said that, while it is difficult and uncommon for inspections alone to identify seafood imports caught with IUU fishing, inspections may lead to further investigation of the documentation that can identify such imports. In addition, inspections can serve as deterrents because they can result in monetary penalties and rejection of high-value goods.

CBP and NMFS Share Information through Several Mechanisms, but NMFS Officials Report Difficulties Obtaining Information from CBP

CBP and NMFS share information through several mechanisms—CBP data systems, direct interaction between officials at ports, and an interagency coordination center—but NMFS officials cited difficulties in efficiently obtaining information from CBP to help combat imports of seafood caught through IUU fishing.

Data systems. CBP's data systems, including the Automated Commercial Environment, hold information submitted by importers or customs brokers about their incoming shipments, including seafood. CBP grants NMFS and other partner agencies access to selected data from these systems, as well as to accompanying analysis tools.[35] One such tool is the Government Client Manifest Capability, which was created in 2019.[36] It provides NMFS and other partner agencies with the ability to query real-time information on imports to support agencies' efforts to target illegal imports, including seafood potentially caught through IUU fishing.

However, NMFS officials told us that this tool has not provided users with the necessary search capabilities to effectively target imports of seafood potentially caught through IUU fishing. Instead, the officials told us that they have had to use less efficient alternative methods that they have developed to retrieve data they need using this tool. NMFS officials told us that these alternative methods are cumbersome and time consuming and do not always work because of system limitations.

---

[35] The information that CBP shares with NMFS is specified in an interagency memorandum of understanding dated 2019.

[36] CBP officials told us that the full name of this system is Automated Targeting System-Government Client Manifest Capability because it is an application of the larger Automated Targeting System. For the purposes of this report, we refer to it as the Government Client Manifest Capability.

According to NMFS officials, NMFS previously had the search capabilities they needed using a CBP tool called the Automated Targeting System- Import Cargo. This tool is used within CBP to identify a broad range of threats. However, CBP officials told us that the agency reviewed the data available to partner government agencies after a system update and determined that CBP needed to provide stronger controls around these agencies' access to sensitive information. As a result, CBP told us they created the Government Client Manifest Capability tool to properly control access to the data. However, until recently, this tool had a technical limitation that has limited agencies' search capabilities, according to CBP officials.

Officials at CBP told us that they recently remedied this search limitation and that the expanded search capabilities have been tested by one partner agency. These officials told us that they notified NMFS and other agencies of the remedy at an interagency meeting in late March 2023.

Direct interaction between officials. CBP and NMFS officials operating at individual ports also share information through direct interaction with one another. For example, NMFS officials at one port described an example of working directly with CBP managers at that port to get information about inbound shipments of imported fresh seafood suspected of being caught through IUU fishing. Because they were able to receive this information in a timely manner, NMFS officials were able to investigate the shipments before they cleared customs. Additionally, these NMFS officials told us that they are members of a regional port working group where they share information with CBP colleagues about ongoing cases of interest, planned operations, and incoming intelligence. These officials told us that gathering intelligence can be challenging but cited examples of local information sharing that provided information that helped them interdict shipments of seafood potentially caught through IUU fishing.

Interagency coordination center. As discussed earlier, CBP's interagency Commercial Targeting and Analysis Center (CTAC) is a mechanism to help partner agencies, including NMFS, coordinate to combat violations of U.S. import laws. CBP officials told us that the center holds weekly meetings in which partner agencies share information about concerns, ongoing operations, and relevant intelligence. NMFS officials told us that CTAC is the primary means through which NMFS shares information on seafood imports with CBP and other agencies with authority over aspects of seafood imports, such as the U.S. Fish and Wildlife Service and Food and Drug Administration.

Officials from CBP and NMFS told us that CBP directs partner agencies to request desired information through CTAC, rather than directly from CBP officials in ports, if CTAC has initiated enforcement efforts, is already generating related intelligence, or is investigating or targeting the shipment of interest. Specifically, these officials told us that in these situations, if NMFS officials in a port need information on a specific shipment, have questions about intelligence, or want to place a hold on a shipment to inspect it, they should reach out to NMFS headquarters, which will reach out to CBP at CTAC to obtain this information. CTAC will provide the information to CBP's Office of Field Operations, which will provide it to NMFS headquarters to share with NMFS port-level officials. Further, CBP officials noted that, in these situations, it is critical for NMFS to also communicate the results of any of its examinations or follow-up activities back to CTAC to ensure targeting data is continually refined.

However, NMFS officials at the headquarters and port levels have expressed concern about the length of time it takes to obtain information through this process. NMFS officials told us that if they want to inspect an incoming shipment, they need sufficient advance notice to coordinate with CBP and to identify officials to conduct the inspection prior to the shipment entering commerce. Consequently, NMFS officials at headquarters and at ports stressed the importance of obtaining timely information to support targeting efforts and said that requesting information through CTAC, in some cases, complicates the process of requesting information. NMFS port officials said that going through CTAC for information was sometimes inefficient and that requesting information through CTAC could take as long as a week.

NMFS officials at ports and at headquarters told us that some information requests are more efficiently addressed by sharing information directly between CBP and NMFS officials operating at ports, rather than making such requests through CTAC, as it currently operates. According to NMFS officials, consultation with CBP counterparts at the port level is important because the import process is complex, and details provided for some imports in CBP's data systems may be subject to interpretation.

Additionally, these officials told us that direct consultation with CBP officials at ports allows NMFS officials to benefit from expertise that CBP personnel may have regarding aspects of the import process or conditions at a particular port.

CBP told us that making information requests through CTAC, rather than through officials at the port level, is important to ensure that officials in CBP headquarters are aware of actions such as requesting information or holds on shipments and how such requests may affect ongoing CBP operations, such as jeopardizing or duplicating them. Additionally, these officials told us that requiring partner agencies to make requests through CTAC can prevent duplication of effort between ports or conflicts among other CBP offices.

The information-sharing process through CTAC, as it currently operates, is not fully consistent with CTAC's stated mission, as specified in the interagency memorandum of understanding. This memorandum establishes how partner government agencies participate in CTAC. CTAC's mission includes maximizing cooperation among participating agencies and facilitating information sharing to combat import violations, including imports of seafood caught through IUU fishing. Nevertheless, NMFS officials told us that CBP's current process may not always provide the information they need in a timely fashion. They said that the amount of time it takes to receive information could undermine time-sensitive efforts to target, investigate, or identify imports of concern. Unless CTAC officials work with NMFS to ensure that it has timely access to information through CTAC on seafood imports that may have been caught with IUU fishing, both NMFS and CBP may be missing opportunities to combat such fishing and associated import violations.

Conclusion

NMFS and CBP both undertake efforts to identify and target for investigation imports of seafood potentially caught through IUU fishing. To support these efforts, the agencies share information through mechanisms, including CBP-managed data and analysis tools and a CBP interagency coordination center, CTAC. However, NMFS officials told us that they experienced difficulties obtaining timely information through CTAC. Such information could help them better identify and target seafood imports for investigation. CBP officials emphasized that coordinating requests for information or holds on shipments through CTAC is important to ensure that officials in CBP headquarters are aware of such requests and how such requests may affect ongoing CBP operations. However, it is important for CBP to balance its need for awareness of information requests with providing timely information to NMFS, as one of its partner agencies. Until CBP works with NMFS to improve the timeliness of its information sharing with NMFS, both NMFS and CBP may miss opportunities to combat imports of seafood caught through IUU fishing.

Recommendation for Executive Action

We are making the following recommendation to CBP:

The Commissioner of CBP should direct relevant officials to work with NMFS to ensure that NMFS has timely access to information it needs to combat imports of seafood caught through IUU fishing. (Recommendation 1)

Agency Comments

We provided a draft of this report to the Departments of Commerce and Homeland Security for review and comment. The Department of Commerce told us that they had no comments on the draft report. The Department of Homeland Security provided technical comments, which we incorporated as appropriate, and written comments reproduced in appendix I. The Department of Homeland Security concurred with our recommendation and said they will take action to address it.

We are sending copies of this report to the appropriate congressional committees, the Secretary of Commerce, the Secretary of Homeland Security, and other interested parties. In addition, the report will be available at no charge on the GAO website at https://www.gao.gov.

If you or your staff have any questions about this report, please contact me at (202) 512-3841 or johnsoncd1@gao.gov. Contact points for our Offices of Congressional Relations and Public Affairs may be found on the last page of this report. GAO staff who made key contributions to this report are listed in appendix II.

Cardell D. Johnson
Director, Natural Resources and Environment

Appendix I: Comments from the Department of Homeland Security

U.S. Department of Homeland Security
Washington, DC 20528

April 28, 2023

Cardell D. Johnson
Director, Natural Resources and Environment
U.S. Government Accountability Office
441 G Street, NW
Washington, DC 20548

Re: Management Response to Draft Report GAO-23-105643, "COMBATING ILLEGAL FISHING: Better Information Sharing Could Enhance U.S. Efforts to Target Seafood Imports for Investigation"

Dear Mr. Johnson:

Thank you for the opportunity to comment on this draft report. The U.S. Department of Homeland Security (DHS or the Department) appreciates the U.S. Government Accountability Office's (GAO) work in planning and conducting its review and issuing this report.

DHS leadership is pleased to note GAO's recognition that U.S. Customs and Border Protection (CBP) and the National Oceanic and Atmospheric Administration (NOAA), National Marine Fisheries Service (NMFS) work to combat imports of seafood caught through illegal, unreported, and unregulated (IUU) fishing, and that CBP and NMFS share information with each other in several ways, including through CBP's Commercial Targeting and Analysis Center (CTAC). CBP remains committed to collaborative information-sharing efforts that help maximize opportunities to combat imports of seafood caught through IUU fishing.

The draft report contained one recommendation for CBP with which the Department concurs. Enclosed find our detailed response to the recommendation. DHS previously submitted technical comments addressing several accuracy, contextual, and other issues under a separate cover for GAO's consideration.

Again, thank you for the opportunity to review and comment on this draft report. Please feel free to contact me if you have any questions. We look forward to working with you again in the future.

Sincerely,

JIM H CRUMPACKER Digitally signed by JIM H CRUMPACKER
Date: 2023.04.28 16:10:09 -04'00'

JIM H. CRUMPACKER, CIA, CFE
Director
Departmental GAO-OIG Liaison Office

Enclosure

Enclosure: Management Response to Recommendation Contained in GAO-23-105643

GAO recommended that the Commissioner of CBP:

Recommendation 1: Direct relevant officials to work with NMFS to ensure NMFS has timely access to information it needs to combat imports of seafood caught through IUU fishing.

Response: Concur. CBP's Office of Trade, CTAC will create a schedule of data that can be shared with NMFS, as well as a proposed timeline on the frequency of sharing this data. The CTAC will also elicit feedback from NMFS to ensure that the data sharing arrangement supports interagency IUU fishing enforcement efforts. Estimated Completion Date: September 29, 2023.

Appendix II: GAO Contact and Staff Acknowledgments

GAO Contact

Cardell D. Johnson, (202) 512-3841 or johnsoncd1@gao.gov.

Staff Acknowledgments

In addition to the contact named above, Elizabeth Erdmann (Assistant Director), Emily Norman (Analyst in Charge), Krista Breen Anderson, Cathleen Carr, Patricia Moye, Cynthia Norris, and Dan C. Royer made key contributions to this report.

GAO's Mission

The Government Accountability Office, the audit, evaluation, and investigative arm of Congress, exists to support Congress in meeting its constitutional responsibilities and to help improve the performance and accountability of the federal government for the American people. GAO examines the use of public funds; evaluates federal programs and policies; and provides analyses, recommendations, and other assistance to help Congress make informed oversight, policy, and funding decisions. GAO's commitment to good government is reflected in its core values of accountability, integrity, and reliability.

Connect with GAO

Connect with GAO on Facebook, Flickr, Twitter, and YouTube. Subscribe to our RSS Feeds or Email Updates. Listen to our Podcasts. Visit GAO on the web at https://www.gao.gov.

To Report Fraud, Waste, and Abuse in Federal Programs

Contact FraudNet:
    Website: https://www.gao.gov/about/what-gao-does/fraudnet Automated answering system: (800) 424-5454 or (202) 512-7700.

Congressional Relations

A. Nicole Clowers, Managing Director, ClowersA@gao.gov, (202) 512-4400, U.S. Government Accountability Office, 441 G Street NW, Room 7125, Washington, DC 20548.

Public Affairs

Chuck Young, Managing Director, youngc1@gao.gov, (202) 512-4800.
U.S. Government Accountability Office, 441 G Street NW, Room 7149 Washington, DC 20548.

Strategic Planning and External Liaison

Stephen J. Sanford, Managing Director, spel@gao.gov, (202) 512-4707.
U.S. Government Accountability Office, 441 G Street NW, Room 7814, Washington, DC 20548.

Chapter 5

National 5-Year Strategy for Combating Illegal, Unreported, and Unregulated Fishing[*]

**Janet Coit
and Richard W. Spinrad**

Member Agencies

    National Oceanic Atmospheric Administration
    U.S. Department of State
    U.S. Coast Guard
    Council on Environmental Quality
    Director of National Intelligence, Representative of National Security Council
    Office of Management and Budget
    Office of Science and Technology Policy
    Office of the U.S. Trade Representative
    U.S. Agency for International Development
    U.S. Department of Agriculture
    U.S. Department of Defense
    U.S. Department of Homeland Security
    U.S. Department of Justice
    U.S. Department of Labor
    U.S. Department of the Treasury
    U.S. Federal Trade Commission

---

[*] This is an edited, reformatted and augmented version of Report to Congress prepared by U.S. Interagency Working Group on IUU Fishing, dated 2022-2026.

In: Ongoing Efforts to Combat Illegal, Unreported …
Editor: Gordon B. Maddox
ISBN: 979-8-89530-858-5
© 2026 Nova Science Publishers, Inc.

U.S. Fish and Wildlife Service
U.S. Food and Drug Administration
U.S. Immigration and Customs Enforcement
U.S. Navy

The National Defense Authorization Act for Fiscal Year 2020 (Public Law No: 116-92, S.1790) Included the Following Language

STRATEGIC PLAN.—Not later than 2 years after the date of the enactment of this title, the Working Group, after consultation with the relevant stakeholders, shall submit to the Committee on Commerce, Science, and Transportation of the Senate, the Committee on Foreign Relations of the Senate, the Committee on Appropriations of the Senate, the Committee on Transportation and Infrastructure of the House of Representatives, the Committee on Natural Resources of the House of Representatives, the Committee on Foreign Affairs of the House of Representatives, and the Committee on Appropriations of the House of Representatives a 5-year integrated strategic plan on combating IUU fishing and enhancing maritime security, including specific strategies with monitoring benchmarks for addressing IUU fishing in priority regions.

I. Executive Summary

The Maritime Security and Fisheries Enforcement Act of December 2019 (Maritime SAFE Act) directed 21 federal agencies to establish an Interagency Working Group on Illegal, Unreported, and Unregulated (IUU) Fishing to serve as the central forum to coordinate and strengthen their efforts to counter IUU fishing and related threats to maritime security. This National Strategy for Combating Illegal, Unreported, and Unregulated Fishing establishes the Working Group's priorities to combat IUU fishing, curtail the global trade in seafood and seafood products derived from IUU fishing, and promote global maritime security. This Strategy provides a framework for coordination for the next five years among the relevant U.S. Government (USG) agencies, in partnership with other governments and authorities, seafood industry, academia, and nongovernmental (NGO) stakeholders that will be key in continuing to make tangible progress in addressing IUU fishing and carrying out a shared vision for stewardship of marine resources. It also supports the objectives outlined in the President's National Security Memorandum on Combating Illegal, Unreported, and Unregulated Fishing and Associated Labor Abuses.

For decades, IUU fishing has been a global problem affecting ocean ecosystems, threatening economic and food security, and putting law-abiding fishermen and seafood producers at a disadvantage. The United States has been a leader in combating IUU fishing and will continue to promote sustainable fisheries management, effective fisheries enforcement, and monitoring of the trade of fish and fish products. We work at regional and international levels, including through the use of key domestic tools, to improve fisheries management and combat IUU fishing on the high seas and within other countries' jurisdictions. This Strategy includes and builds on existing activities with new initiatives to form a comprehensive set of actions to address IUU fishing and associated forced labor, including preventing importation of IUU fish and fish products or those associated with forced labor.

The Maritime SAFE Act charged the Working Group with developing priority regions and priority flag states to be the focus of assistance coordinated by the Working Group. The Act defines "priority regions" as those "at high risk for IUU fishing activity or the entry of illegally caught seafood into the markets of countries in the region; and in which countries lack the capacity to fully address the illegal activity." A priority flag state is defined as a country whose vessels are "actively engaged in, knowingly profit from, or are complicit in IUU fishing" and, at the same time, "is willing, but lacks the capacity, to monitor or take effective enforcement action against its fleet."

Through the Working Group, the United States aims to employ a coordinated, cohesive, and regionally appropriate approach to combating IUU fishing and related threats to maritime security in priority regions and priority flag states. The Working Group developed a comprehensive, tiered list of 12 priority regions intended to focus USG capacity building, training, and information sharing efforts to combat IUU fishing and enhance maritime security. Within these priority regions, the Working Group selected five priority flag states and administrations with which to pursue new projects and initiatives to support ongoing counter- IUU fishing efforts: Ecuador, Panama, Senegal, Taiwan, and Vietnam. These five flag states or administrations were not selected as priorities because they are the worst IUU fishing offenders. Rather, each flag state or administration has demonstrated a willingness to and interest in taking effective action against IUU fishing activities associated with its vessels. The United States aims to assist these governments and authorities to become self-sufficient, regional leaders in the fight against IUU fishing. The United States will engage with additional flag states and continue in this manner with the goal of building a coalition across the globe that works in tandem to prevent and eliminate IUU fishing.

The United States will work to build positive, cooperative partnerships globally, but particularly within the priority regions and with the priority flag states and administrations to support and enhance both ongoing and new counter-IUU fishing efforts. These efforts are organized under three strategic objectives we are advancing to combat IUU fishing:

1) Promote Sustainable Fisheries Management and Governance – We will continue to collaborate with other countries, authorities, and stakeholders and within international and regional organizations to ensure fisheries management and conservation practices are effective and science-based. We will advance well-managed fisheries that support the food security, livelihoods, and sustainable development of communities around the world, particularly in developing countries. Through capacity building and other efforts, we will promote comprehensive fisheries management and governance structures, including enforcement mechanisms with robust and effective laws, to encourage compliance and guard against IUU fishing.

2) *Enhance the Monitoring, Control, and Surveillance of Marine Fishing Operations* – We will continue aggressively conducting at-sea monitoring, control, and surveillance operations to compel compliance with U.S. law and international fisheries conservation and management measures. We will improve coordination and information sharing across intelligence, enforcement, and regulatory agencies, as well as foreign partners and non-governmental organizations (NGOs), to increase enforcement effectiveness. We will prioritize enforcement efforts against IUU fishing networks, including transnational criminal organizations. We will promote the adoption of at-sea enforcement provisions in multilateral agreements. We will help improve global enforcement efforts by increasing partner capacity and through coordinated operations with like-minded nations.

3) *Ensure Only Legal, Sustainable, and Responsibly Harvested Seafood Enters Trade* – We will use our position as a major seafood-consuming and fishing nation to support collective efforts to ensure the effective conservation and management of international fisheries. We will work to ensure that imported seafood is safe, legal, and comes from sustainably managed fisheries. We will also work to identify and address labor abuses, including forced labor, throughout the seafood supply chain.

This Strategy is a product of a series of interagency discussions among the Working Group members, including planning sessions that focused on USG work that is underway or planned for enhancing maritime security and combating IUU fishing in the priority regions and flag states and administrations. It was also informed by comments from NGOs and industry associations submitted in fall 2021. In this Strategy, we emphasize a broadening of USG efforts and enhanced coordination and collaboration across the Working Group agencies, as well as collaboration with foreign partners and nongovernmental organizations. For each of the three strategic objectives, we identified a wide range of actions that can build on existing efforts or that we plan to initiate in the next five years. These include specific programs that focus on particular countries, territories, and entities, in priority regions and flag states as we seek to target our work and build over time a global network that can support broad local, regional, and international efforts to combat IUU fishing. We also include efforts that the USG continues to undertake at a regional or international level, as well as identify the key domestic tools the United States uses to bring about improvements in fisheries management in other countries or prevent products of IUU fishing from being imported into the United States. Finally, we identify monitoring benchmarks for the strategic objectives to help assess progress in the priority regions and priority flag states and administrations. Progress on this Strategy within the five-year period will depend on the availability of resources (both in terms of funding and personnel) of all relevant agencies, as well as engagement by key private sector partners, partner governments and authorities, and support from U.S. stakeholders.

II. Introduction

IUU fishing is a global problem that threatens ocean ecosystems and sustainable fisheries. It also threatens our economic security and the natural resources that are critical to global food security, deprives scientists of data needed to inform sound fisheries conservation and management decisions, and puts law-abiding fishermen and seafood producers in the United States and abroad at a disadvantage.

Over the past two decades, the United States has been a leader in building an expansive toolbox for countries and partners—individually and collectively—to address IUU fishing, bringing world-wide recognition to the issue through international fora and making progress through major domestic initiatives such as the National Plan of Action of the United States of America to Prevent, Deter, and Eliminate Illegal, Unregulated, and Unreported Fishing (NPOA-IUU)[1] and the Presidential Task Force on Combating IUU Fishing and Seafood Fraud. Through this work, we have improved our knowledge of how we, as the different components of the USG, need to work collaboratively to understand and address IUU fishing. We recognize even more clearly that addressing IUU fishing is not just about fish: it is a multi-faceted problem that covers other core policy concerns, including human rights, food security, and maritime security.

Our improved knowledge and understanding has prepared us to fulfill a new mandate, the Maritime SAFE Act, which became law on December 20, 2019 and supports a "whole-of- government approach across the Federal Government to counter IUU fishing and related threats to maritime security." It seeks to achieve this through a number of means, including improving data sharing that enhances surveillance, advancing effective enforcement and prosecution against IUU fishing, supporting coordination and collaboration, increasing and improving global transparency and traceability across the seafood supply chain, responding to poor working conditions and labor abuses in the fishing industry, improving global enforcement operations, and preventing the use of IUU fishing as a financing source for transnational organized groups.

Part II of the Maritime SAFE Act establishes a collaborative interagency Working Group to strengthen maritime security and combat IUU fishing. In June 2020, this group met for the first time (see Appendix C for additional details about the Working Group's leadership and membership). The Working

---

[1] https://2001-2009.state.gov/documents/organization/43101.pdf.

Group created the following subgroups to carry out specific activities to fulfill the Working Group's mandated responsibilities:

- Maritime Intelligence Coordination
- Public-Private Partnerships
- Labor in the Seafood Supply Chain
- Gulf of Mexico IUU Fishing
- Priority Regions and Flag States

As a part of the process for developing this Strategy, the Working Group first identified the priority regions and priority flag states, as defined under section 3552(b) of the Maritime SAFE Act. To select the priority regions, the Working Group assessed different regions around the world through a framework that considered information about recent or egregious cases of IUU fishing in each region; the diverse coastal countries', territories', and entities' institutional and operational capacity to deal with IUU fishing; and how the countries, territories, and entities within the region participate in the seafood supply chain and global markets. Based on this analysis, the Working Group developed a comprehensive, tiered list of 12 priority regions intended to focus USG capacity building, training, and information sharing efforts to combat IUU fishing and enhance maritime security (see Appendix A for the priority regions).

With an understanding of the global picture of IUU fishing, the Working Group evaluated several additional criteria to better understand the IUU fishing problem in each of the flag states and administrations within these 12 priority regions, including IUU fishing activities by their vessels, their institutional and operational capacity to monitor and police their fleets and exclusive economic zones (EEZs) effectively, and their willingness to work with the United States to address IUU fishing activities by their vessels. Based on this analysis, the Working Group selected the following five priority flag states and administrations from the priority regions with which to pursue projects and initiatives to support counter-IUU fishing efforts: Ecuador, Panama, Senegal, Taiwan, and Vietnam (see Appendix B). These five were not selected as priorities because they are the worst IUU fishing offenders. Rather, each has demonstrated a willingness and interest to take effective action against IUU fishing activities associated with its vessels. Ongoing and new counter-IUU fishing efforts will be supported and enhanced by a positive, cooperative partnership with the United States. The relationships with these flag states and administrations are expected to lead directly to meaningful progress on combating IUU fishing, as well as support future partnerships with other countries and partners that expand the reach of U.S. efforts.

The United States has established the following three strategic objectives to combat IUU fishing and to address related maritime security threats at a global level, with specific efforts under each objective that are focused in the priority regions and flag states and administrations:

1) Promote sustainable fisheries management and governance.
2) Enhance monitoring, control, and surveillance of marine fishing operations.
3) Ensure only legal, sustainable, and responsibly harvested seafood enters trade.

Over the next 5 years, the United States will use these strategic objectives to develop projects and initiatives to combat IUU fishing within the priority regions and in each of the priority flag states and administrations. U.S. activities will be tailored to both the specific needs of each region or flag state or administration and the range of U.S. projects and activities already underway. We will also make full use of existing domestic tools, technologies, and our ongoing efforts to advance guidance, conservation and management measures, and cooperation in international fora to achieve these strategic objectives. Monitoring benchmarks are identified within each strategic objective, which were designed to track progress in combating IUU fishing and enhancing maritime security, particularly in the priority regions and flag states and administrations.

The efforts of the Working Group align closely with the President's National Security Memorandum (NSM) on Combating Illegal, Unreported, and Unregulated Fishing and Associated Labor Abuses. Specifically, the NSM directs agencies to increase coordination among themselves and with diverse stakeholders to work towards ending forced labor and other crimes or abuses in IUU fishing; promote sustainable use of the oceans in partnership with other nations and the private sector; and advance foreign and trade policies that benefit U.S. seafood workers.

III. Strategic Objective 1: Promote Sustainable Fisheries Management and Governance

The foundation for combating IUU fishing is ensuring that coastal and flag states have processes and frameworks in place to manage the fish stocks they harvest sustainably. Because many fish and other marine wildlife cross national boundaries, the manner in which other countries and authorities manage shared marine resources can directly affect the status of fish stocks and protected or endangered species of importance to the United States. The first step is ensuring that, individually or collectively, countries and authorities have put in place the basic policies, laws, and regulations for the responsible conservation and management of fisheries resources.

This requires multifaceted efforts, including determining and monitoring the biological status of stocks, implementing science-based harvest limits and accountability measures, minimizing bycatch, providing for adequate enforcement, and combining these efforts to implement ecosystem-based fisheries management. Just because a fish stock is harvested legally may not mean that it is harvested sustainably, and enforcement is only meaningful when there are effective laws and regulations in place to begin with. We will undertake a range of actions to accomplish this strategic objective. For some of these actions, we identify the country or administration and the type of work, but we will also seek to implement them in additional countries, where appropriate, as resources become available.

In Priority Regions and Flag States or Administrations

1) In partnership with academia and NGOs, assess the scientific, legal, regulatory, and operational systems of the countries and authorities in the priority regions for managing marine fisheries to understand their gaps and needs and where and how to focus technical assistance.
    a) Recent public attention and U.S. initiatives have focused on distant-water fishing off the coastal states of South America, particularly around Ecuador. We will organize periodic meetings to bring together U.S. entities involved in countering IUU fishing in or near Ecuador, including agencies, NGOs, and academia, to share information, successes, and failures; find areas of coordination; and avoid duplicating efforts. Such meetings will

include the government and other sectors of Ecuador as appropriate. In addition, the United States is entering a memorandum of understanding (MOU) for collaboration in the Eastern Tropical Pacific Marine Corridor (CMAR) with the governments of Colombia, Ecuador, Panama, and Costa Rica.

    b) We will work with Panama, interested parties in academia, and NGOs to assess the fishing (harvesting and carrier) vessel monitoring and control systems within Panama and use the assessment to identify ways to help build the governance and operational capacity of Panama.

    c) Vietnam has demonstrated a commitment to improving its fisheries management and governance structures despite challenges in policy coordination between its central and regional governments. We will work with Vietnam to help develop coordination mechanisms modeled on U.S. regional fishery management councils to achieve better alignment and buy-in across all levels of government.

    d) We will aim to replicate a recent Department of State Bureau of International Narcotics and Law Enforcement project in the Caribbean, Central America, and South America regions that provides baseline information on gaps in understanding around country law enforcement and legislative capacity on IUU fishing. The current project analyzes the scope of IUU fishing–related legislation in the region, identifies government-backed malign actors perpetrating IUU fishing, assesses crimes associated with IUU fishing, and collects local anecdotes on the adverse impacts of IUU fishing in the region. Current priority countries include Costa Rica, Ecuador, Panama, Guyana, Jamaica, Suriname, Chile, Argentina, and Uruguay.

2) Enhance U.S. efforts to improve the capacity of coastal states and authorities to manage their domestic fisheries and to combat IUU fishing, both by building the political will to devote resources to these issues and by providing information, equipment, and expertise.

    a) The United States and Taiwan have committed to explore opportunities to conduct combined security and law enforcement operations in the Indo-Pacific region through an MOU between the Taipei Economic and Cultural Representative Office in the United States and the American Institute in Taiwan. Under this MOU, we will increase professional exchanges between Taiwan

Coast Guard and U.S. Coast Guard law enforcement officers; explore similarities between law enforcement teams, authorities, and jurisdictions; and jointly contribute to combating the increasingly complex transnational IUU fishing problem in the Indo-Pacific region.

b) We will help to improve fisheries management in Senegal through the U.S. Agency for International Development (USAID) Feed the Future Dekkal Geej (DG) project. The five-year $15 million project is improving the sustainability of Senegal's artisanal fisheries sector with an emphasis on co-management practices, data collection and analysis, policy support, food security, and resiliency. Additionally, the DG project helps Senegalese stakeholders from the national to the village level to detect and deter IUU fishing.

3) Develop best practices for foreign licensing and recommendations on the contents of a model fishing access agreement to ensure adequate oversight, transparency, and accountability of foreign vessels. Robust access agreements and the capacity to monitor them minimize the risk of IUU fishing and overexploitation by distant-water fleets operating in coastal state EEZs. In the Tier 1 and Tier 2 priority regions, we will work with coastal states and authorities to apply the best practices and model access agreements to their respective EEZs by providing advice on laws and policies that will enable effective governance and law enforcement in their EEZs. We will also encourage nations to demarche the flag states that have foreign fishing fleets that violate fisheries laws and regulations in their territorial seas and EEZs.

4) Partner with NGOs, media outlets, and academia to promote sound, fact-based journalism and build journalistic capacity to report and educate the public on the importance of sustainable fisheries management, expose known instances and actors involved in IUU fishing, and highlight any clear and actionable intersections with distant water fishing, flags of convenience, and forced labor.

5) Encourage nations in priority regions and priority flag states and administrations to adopt, or update any existing, National Plan of Action on counter–IUU fishing and update our own to serve as a model.

At the International and Regional Levels

1) Continue championing within regional fisheries management organizations (RFMOs) effective, science-based conservation and management measures and stronger and more transparent compliance schemes.
    a) RFMOs have put in place a range of conservation and management measures intended to support the long-term sustainability of the resources under their mandates, as well as monitoring, control, and surveillance (MCS) measures to ensure compliance with these rules. We will press for RFMOs to fill any gaps in management and MCS measures, or to strengthen and enhance those already in place, particularly measures for fisheries data collection and reporting, monitoring and controlling transshipment, mitigating bycatch of protected species, tracking trade of fish and fish products, and enabling high seas boarding and inspection. We will also work to ensure that RFMO port state measures schemes are consistent with the Agreement on Port State Measures (PSMA). Where there are gaps in management of particular species, we will seek to have the relevant RFMO address the gaps or, where an RFMO is lacking, press for the formation of a management body.
2) Advance policies and programs through the Food and Agriculture Organization of the United Nations (FAO) that promote ecosystem-based fisheries management and ensure the long-term sustainability of fisheries, both small-scale and industrial. Continue to lead the development of voluntary and binding instruments that strengthen global fisheries management.
    a) Champion the implementation of the international guidelines on transshipment recently developed by FAO that outline best practices and help to increase monitoring to close loopholes that facilitate IUU fishing.
    b) Continue to work toward full implementation of the Asia-Pacific Economic Cooperation's (APEC) Roadmap to Combat IUU Fishing through robust participation in the Oceans and Fisheries Working Group (OFWG) of APEC. Participate in the inter-sessional working group that OFWG established to address current gaps in the implementation plan. Furthermore, the United States will continue to demonstrate leadership on IUU fishing in

APEC through the hosting and co-sponsorship of projects that fall within the purview of the Roadmap to Combat IUU Fishing. This leadership will be even more on display in 2023, when the United States will be the APEC host country.

Use of Existing Domestic Tools

1) Continue implementing the international provisions of the High Seas Driftnet Fishing Moratorium Protection Act.[2]
2) Continue implementing the international provisions of the Marine Mammal Protection Act.[3]

Monitoring Benchmarks: To assess progress in fisheries management frameworks within priority regions, we will evaluate the extent of improvements in fisheries governance of coastal states and authorities in the priority regions, including new fisheries laws, fisheries management plans, application of effective MCS mechanisms, and the mechanisms for taking enforcement actions. To assess priority flag states' and administrations' progress in controlling their respective distant water fleets, we will examine, where applicable, their improvements in fisheries governance, with particular focus on new laws and application of effective MCS mechanisms. We will select specific metrics for assessing progress in a given priority region or priority flag state or administration that take into account the unique needs and circumstances within each, including the variability in fisheries governance frameworks. These metrics will draw from the following list, which may also be augmented by additional metrics appropriate to a given priority region or flag state or administration:

- Accession to and/or progress made in implementing international instruments, including, but not limited to: PSMA,[4] NPOAs on combating IUU fishing, Regional Plans of Action to Promote Responsible Fishing Practices, the Global Record of Fishing Vessels,

---

[2] https://www.fisheries.noaa.gov/international/international-affairs/report-iuu-fishing-bycatch-and-shark-catch.3
[3] https://www.fisheries.noaa.gov/foreign/marine-mammal-protection/noaa-fisheries-establish es -international- marine-mammal-bycatch-criteria-us-imports.
[4] https://www.fao.org/port-state-measures/en/.

Refrigerated Transport Vessels and Supply Vessels (Global Record),[5] and the Agreement to Promote Compliance with International Conservation and Management Measures by Fishing Vessels on the High Seas (Compliance Agreement).[6]

- Increase in the number of countries joining or cooperating in bilateral and/or regional fisheries management agreements, and bilateral or multilateral IUU fishing enforcement agreements.
- Increase in the uptake and implementation of vessel monitoring system (VMS) programs, centralized VMS reporting, and/or other vessel monitoring systems.
- Increase in the availability of information regarding the proportion of foreign-owned fleets and/or flag vessels operating in EEZs (e.g., fishing access agreements, reports of incursions, vessels flagged to states with known IUU fishing concerns).

In addition to the above quantitative metrics, we will also use qualitative assessment questions, such as:

- Have countries implemented or expressed an intention to join the PSMA?
- Did U.S. activities result in new (or changes to existing) fishing access agreements so that key coastal states and authorities have better oversight, transparency, and accountability of foreign vessels?
- Did support from the United States result in improvements in the regulation and monitoring of transshipment activities in the Tier 1 and Tier 2 priority regions (such as establishment of a transshipment monitoring regime with observers, pre-notification, post-transshipment reporting, and VMS)?

We will also assess progress made within RFMOs to strengthen conservation and management measures essential to sustainable fisheries management, such as comprehensive MCS measures, bycatch mitigation measures, and trade-tracking mechanisms. Finally, we will take into account fisheries management and enforcement improvements made within countries identified under the High Seas Driftnet Fishing Moratorium Protection Act or

---

[5] https://www.fao.org/global-record/en/.
[6] https://www.fao.org/iuu-fishing/international-framework/fao-compliance-agreement/en/.

improvements in bycatch mitigation measures brought about by the import provisions of the Marine Mammal Protection Act.

IV. Strategic Objective 2: Enhance the Monitoring, Control, and Surveillance of Marine Fishing Operations

In the vast open spaces of the world's oceans, MCS of marine fishing operations is critical to ensure compliance with national and international fishing rules, regulations, and conventions. Without effective MCS programs, there is little deterrence against IUU fishing. We will improve MCS coordination within the United States and with our international partners to increase the risks and costs of IUU fishing, hold flag states and administrations accountable for the actions of their fishing fleets, and improve the capacity of countries in priority regions to protect their respective sovereignty. We will undertake a range of actions to accomplish this strategic objective.

In Priority Regions and Flag States and Administrations

1) Collaborate and coordinate with foreign navies, coast guards, and other enforcement agencies to optimize our collective capacity to design and implement effective MCS programs and improve the process for receiving and responding to reports of illicit activity, investigations, and the legal infrastructure to adjudicate IUU fishing cases.
    a) Prioritize fisheries enforcement trainings in priority regions and flag states or administrations to build comprehensive enforcement programs, including any necessary legal mechanisms, for conducting investigations, the use of information from whistleblowers, and electronic evidence collection and computer forensics techniques.
    b) Deliver training to improve the capability of partner law enforcement personnel involved in complex investigations related to IUU fishing that include international matters including demarches, financial issues, government corruption, and other enforcement actions.

c) Continue to leverage the Code of Conduct Concerning the Repressing of Piracy, Armed Robbery against Ships, and illicit Maritime Activity in West and Central Africa (Yaoundé Code of Conduct)[7] to address fisheries governance issues in the Gulf of Guinea region and seek opportunities to replicate the Yaoundé Code of Conduct and associated African Maritime Law Enforcement Partnership (AMLEP)[8] phased trainings, exercises, and legal interoperability studies, as well as regional information sharing centers, in other Tier 1 as well as Tier 2 priority regions.

d) Continue to support ongoing work that focuses on information gathering and law enforcement and legislation/law-making capacity building in the maritime security space, including building capacity to address crimes associated with IUU fishing.

2) Prioritize establishing new bilateral enforcement agreements, commonly referred to as "shiprider agreements," with key countries and authorities in Tier 1 and Tier 2 priority regions, as well as with priority flag states and administrations. Similarly, add counter- IUU fishing provisions to existing shiprider agreements.

a) The existing U.S.-Senegal bilateral enforcement agreement presents an opportunity for the United States to partner with a priority flag state to demonstrate commitment in the Tier 1 priority region of the Gulf of Guinea where IUU fishing has historically occurred relatively unopposed. The United States will find opportunities to use this bilateral agreement to help protect Senegalese fisheries resources and, in turn, display to the region the value of collaboration with the United States in the fight against IUU fishing.

3) Continue to seek opportunities to include counter–IUU fishing aspects in multilateral at- sea exercises with partner navies and coast guards.

a) Maximize opportunities to conduct multi-agency training sessions to reinforce the United States' position on combating IUU fishing and translate best practices across partner agencies. This effort will include synchronizing IUU fishing training

---

[7] https://wwwcdn.imo.org/localresources/en/OurWork/Security/Documents/code_of_conduct%20signed%20from%20 ECOWAS%20site.pdf.

[8] https://www.africom.mil/what-we-do/security-cooperation/africa-maritime-law-enforcement-partnership-amlep-program#:~:text=The%20African%20Maritime%20Law%20Enforcement,combined%20maritime%20law%20enfor cement%20operations.

materials among agencies to form a consistent, whole-of-government message when engaged in capacity building operations.
b) Incorporate counter–IUU fishing subject matter experts and relevant personnel from partner agencies into the multinational, security-focused, multi-dimensional exercises that take place globally.
c) Incorporate counter–IUU fishing subject matter experts and relevant personnel from partner agencies into the multinational Maritime Training Activities (MTA), for example, as was done for the 2021 MTA exercise Sama between the Philippines, Japan, France, and the United States. Look for opportunities to share complex at-sea training to strengthen multilateral forces' ability to work together, building partnerships and interoperability between foreign navy and coast guard ships and aircraft.

To Leverage Technologies and Information Sharing

1) Improve information sharing within and among our foreign partners by building a strong international network to share information that supports maritime enforcement efforts.
   a) Pursue the use of existing multilateral organizations—such as RFMOs, regional coast guard forums, UN Office on Drugs and Crime (UNODC), and the International Criminal Police Organization (INTERPOL)—as platforms for sharing information for maritime enforcement.
   b) Develop standards and, where necessary, bilateral and multilateral agreements to advance information sharing.
   c) Promote the Maritime Information Sharing (MIS) initiatives led by the National Maritime Intelligence-Integration Office (NMIO) to develop unclassified collaborative data sharing processes and capabilities that enhance regional maritime domain awareness and maritime security. Also, build consensus in the interagency on the best U.S.-sponsored IT systems and tools for better uniformity and consistency in international collaboration efforts worldwide.
   d) Use the Department of Defense's Maritime Security Initiative to enhance maritime detection capabilities of eligible countries

within their EEZs, increase maritime domain awareness in Tier 1 and Tier 2 priority regions, and develop a common operating picture for regional information sharing.
2) Establish and improve the legal and infrastructure frameworks to allow for sharing of information and data with civil society, NGOs, and the private sector, as appropriate, on IUU fishing and other connected transnational organized illegal activities, so that we can better leverage external resources. This could involve setting up memoranda of understanding or information sharing agreements with particular entities to be able to share information, as appropriate.

To Enhance Coordination and Collaboration Within the USG

1) Maintain IUU fishing as an intelligence collection priority within the U.S. intelligence community. Intelligence collection efforts will focus on vessels and economic activity connected to transnational criminal networks engaged in IUU fishing and associated threats to maritime security including piracy, narcotics smuggling, and human trafficking, including forced labor.
2) Coordinate law enforcement operations and investigations across U.S. federal agencies, combining unique capabilities, resources, and authorities into a whole-of-government effort to identify, track, and target IUU fishing operations and actors most effectively and efficiently. Coordinate and focus Department of Defense, Department of State, Coast Guard, and USAID international capacity building authorities and resources on counter– IUU fishing priority outcomes and objectives. Make use of established region-specific working groups on IUU fishing and maritime domain awareness to interface regularly with regional partners—such as the working groups hosted by Combatant Commanders—and thereby synchronize to make standardized and more efficient engagement with our partners on these topics.

Monitoring Benchmarks: To assess progress in combating IUU fishing in priority regions and deterring IUU fishing by vessels of priority flag states and administrations, we will evaluate whether or how the capacity to monitor and take enforcement action has improved due to USG efforts. In general, we will assess improvements in MCS and enforcement capacity through the mechanisms for information sharing established or improved upon within a region, improvements in enforcement cooperation, establishment of multilateral enforcement arrangements, and reduced incidences of IUU fishing and crimes associated with IUU fishing. We will select specific metrics for assessing progress in a given priority region or priority flag state or administration that take into account the unique needs and circumstances within each, including the variability in MCS and enforcement capacity among them. These metrics will draw from the following list, which may also be augmented by additional metrics appropriate to a given priority region or priority flag state or administration:

- Increase in the number of bilateral enforcement agreements and similar regional cooperation arrangements related to IUU fishing and maritime security.
- Increase in the number of partner-led counter–IUU fishing operations, particularly in priority regions and flag states or administrations.
- Increase in partner countries' and authorities' patrol assets deployed in support of USG- led operations.
- Increase in the number of information sharing agreements with partner countries and authorities in priority regions.
- Increase in the regularity of MCS and other fisheries enforcement training events conducted in priority regions and with priority flag states or administrations.
- Increase in the frequency of joint agency training sessions to strategic partners.

In addition to the above examples of quantitative metrics, we will also develop qualitative assessment questions such as:

- Were USG foreign engagements appropriately focused in priority regions and with priority flag states or administrations as directed by this Strategy for combating IUU fishing?
- Did the USG, in coordination with INTERPOL and other enforcement bodies, generate actionable intelligence, information, or analysis that effectively identified vessels, flag states or administrations, beneficial owners, or criminal organizations engaged in IUU fishing?

- Did the USG disseminate unclassified systems for the sharing of information with top tier foreign partners or nongovernmental organizations, with built-in collaborative analytic tools?
- Did USG-generated intelligence, information, or analysis effectively support the identification of vessels, flag states or administrations, beneficial owners, or transnational criminal organizations engaged in IUU fishing?
- Did USG-generated intelligence, information, or analysis effectively support U.S. investigations against IUU fishing networks or perpetrators?
- Did USG-generated intelligence, information, or analysis result in foreign port access denials levied against IUU fishing networks or perpetrators?
- Did the USG improve the capability of partner countries or authorities or entities to receive and disseminate IUU fishing–related information?
- Did the USG increase authorities and opportunities to conduct at-sea enforcement operations in areas where there is no applicable RFMO high seas boarding and inspection scheme?
- Did USG operations and reporting result in meaningful sanctions against IUU fishing networks and perpetrators?

V. Strategic Objective 3: Ensure Only Legal, Sustainable, and Responsibly Harvested Seafood Enters Trade

The seafood sector plays an important role in the U.S. economy, generating approximately 1.5 million jobs and providing a nutritious source of protein to the American public and elsewhere around the world. Trade in this sector is also vital, as the United States is one of the largest importers and exporters of seafood in the world. The United States currently imports approximately 80 percent of the seafood it consumes.

Because the United States is a large seafood-consuming, importing, and fishing nation, we must take an active role in shaping the conservation and management regimes of fisheries across the globe. We will continue to work to meet U.S. consumer demand for imported seafood that is safe, legal, responsibly harvested, and sustainable. IUU fishing undercuts the competitiveness of the U.S. seafood sectors that operate in some of the most sustainably managed and heavily regulated fisheries in the world. We address these challenges by engaging other countries and authorities internationally, both directly and through various multilateral organizations.

Within Priority Regions and Flag States or Administrations

1) Continue to assist countries and authorities in working toward adopting and implementing the PSMA and help implement programs that support robust port state measures and capacity to prevent IUU fishing products from entering global seafood markets.
   a) Work with and support PSMA Parties through the existing PSMA sub-bodies, including the Part 6 Working Group, which focuses on and provides funding for capacity building for PSMA Parties, and the Technical Working Group on Information Exchange, which will make key information sharing provisions of the PSMA possible.
   b) Support capacity building through trainings conducted by NOAA in partner countries and authorities to build legal and operational capacity to implement the PSMA, including with Indonesia, Philippines, Thailand, Vietnam, Colombia, Ecuador, and Peru.
2) Engage with key governments and authorities, industry, and NGOs to develop actions to improve transparency of seafood supply chains, including implementing new or strengthening existing traceability schemes and developing related RFMO trade-tracking measures where needed. For traceability schemes, we will promote the use of standardized, secure, electronic data throughout the seafood supply chain for managing chain-of-custody, facilitating compliance with legal requirements, and improving enforcement cases. We will also promote data sharing across traceability schemes and international databases, including standardization of key data elements that are collected as a part of these schemes.
3) Work to increase knowledge within governments and industry about U.S. transparency and traceability standards for imports of seafood and seafood products.
4) Initiate discussions with countries and authorities, bringing in any relevant local or regional coordination bodies, as well as academia and NGOs, to identify labor abuses and human trafficking, including

forced labor concerns in the fishing industry, and to address those concerns, including any technical assistance that could be provided.

At the Regional and International Levels

5) Continue to identify and address poor working conditions and labor abuses.
   a) Initiate discussions on labor standards in the seafood supply chain within international organizations, such as RFMOs and FAO, and, as appropriate, advance measures to address any identified cases of labor abuses and human trafficking, including forced labor.
   b) Engage the FAO, International Labor Organization (ILO), and the International Maritime Organization (IMO) Joint Working Group on IUU Fishing, among others, to promote decent work and advance initiatives to identify and deter forced labor across each of the three organizations.
6) Facilitate public-private partnerships with civil society, including industry, to support efforts to prevent labor abuses in the recruitment phases of the seafood supply chain, as well as efforts to improve data collection and reporting on potential labor abuses aboard vessels and in seafood processing facilities.
7) Continue to advocate at the World Trade Organization for comprehensive and effective disciplines that prohibit certain forms of fisheries subsidies that contribute to overcapacity and overfishing. We will also continue to pursue additional transparency with respect to the use of forced labor on fishing vessels.
8) Convene seafood traceability workshops for stakeholders, inviting a diverse array of experts from a range of countries and authorities, to identify pragmatic and tangible ways to optimize, and coordinate across, traceability schemes and apply the outputs to traceability schemes within the priority regions.
9) Assess financial flows and the use of financial institutions to launder profits related to IUU fishing.

Use of Existing Domestic Tools

10) Under the Seafood Import Monitoring Program,[9] develop a modernized information- systems approach to identify shipments most at risk of containing IUU fish and fish products using predictive analytics, artificial intelligence, machine learning, and cloud technology.
11) Continue to use existing and future trade agreements (e.g., the U.S.-Panama Trade Promotion Agreement), including environmental cooperation agreements and work plans to combat IUU fishing and address forced labor in fishing.
12) Expand and enhance the coordinated use within the USG of Customs and Border Protection's (CBP) Commercial Targeting and Analysis Center (CTAC) and Agriculture and Prepared Products Center of Excellence to take enforcement action at ports that target IUU fish and fish products; for products that may have improper documentation, mislabeling, or suspect origins; and conduct trade security operations that prevent illegally harvested seafood from entering U.S. commerce.
13) Continue the use of intelligence on seafood trade in the priority regions and flag states generated from the Global Agricultural Information Network (GAIN), U.S. Department of Agriculture's (USDA) market intelligence reporting system, and the reports from USDA's Foreign Agricultural Service officers to stay abreast of developments, as information becomes available that are relevant to IUU fishing or this Strategy, and identify responsive actions as needed.
14) Establish additional ad hoc mechanisms, as warranted, to coordinate investigations of the sources, trade flows, seafood labeling, or advertising/marketing of seafood products with a high vulnerability to IUU fishing or seafood fraud and refer the results to the appropriate enforcement agencies, such as NOAA, Coast Guard, CBP, Food and Drug Administration, or Federal Trade Commission. Such coordination could also include private entities, as appropriate.
15) CBP, Coast Guard, and NOAA will develop collaborative mechanisms to share timely information on IUU fishing and its

---

[9] https://www.fisheries.noaa.gov/international/seafood-import-monitoring-program.

relationship to human trafficking and other illicit activities, to enhance coordination to address these illicit activities.
16) Collaborate among the members of the IUU Fishing Working Group with the relevant competences to share timely information on transnational organized illegal activity, including human trafficking and illegal trade in narcotics and arms that may be tied to IUU fishing.

Monitoring Benchmarks: To assess progress in ensuring that only legal, sustainable, and responsibly harvested seafood enters trade, we will evaluate the increase in availability or implementation of trade monitoring and controls with catch documents or certification programs, primarily within the priority regions. In addition, steps in developing and coordinating traceability or catch documentation programs adopted in other key seafood markets (e.g., Japan and the European Union) could also contribute to progress, as these markets can drive improvements in the countries and authorities in the priority regions. We will also examine how international fisheries organizations and other international organizations are contributing at the global level to improving working conditions and ending labor abuses in the seafood supply chain. We will select specific metrics for assessing progress in a given priority region or priority flag state or administration that take into account the unique needs and circumstances within each, including the variability in the status of seafood trade monitoring frameworks. These metrics will draw from the following list, which may be augmented by additional metrics appropriate to a given priority region or priority flag state or administration, some of which may be informed from the textile, logging, or mining sectors:

- Increase in dialogues within priority regions where governments, authorities, industry, and NGOs can discuss establishing traceability schemes.
- Increase in trade monitoring and/or controls with catch documents or certification programs as well as increased or improved interoperability across multiple programs.
- Increase or improvements in implementing PSMA.
- Increase in trade monitoring programs with procedures that examine reports of potential labor abuses aboard vessels and in processing facilities.

- Increase in the availability of processes for data collection and reporting on potential labor abuses aboard vessels and in processing facilities.
- Increase in the number of conservation and management measures, programs, or instruments addressing labor standards in the seafood supply chain adopted by international organizations, such as RFMOs, FAO, and ILO.

In addition to the above quantitative metrics, we will also develop qualitative assessment questions, such as:

- Did internal USG information sharing and coordination result in preventing the entry of products that are associated with IUU fishing into the United States?
- Did USG-generated intelligence, information, and analysis result in economic sanctions in the United States (i.e., CBP's Withhold Release Orders and NOAA seafood import restrictions)?
- Did implementation of U.S. trade agreements bring about progress in achieving this strategic objective?

VI. Conclusion

Through this Strategy, we are reinforcing our efforts to combat IUU fishing so that marine ecosystems are conserved and the food security, livelihoods, and sustainable development of communities around the world are supported. Our approach is to focus now on the identified priority regions and flag states and administrations with an intent to shift to other geographic regions and flag states or administrations as we build the coalition that works in harmony to stamp out IUU fishing. The IUU Fishing Working Group will take stock of progress regularly, using the monitoring benchmarks tailored to the assessments of progress in particular areas or countries and authorities. Such assessments may determine where or how we should adjust a course of action to achieve the strategic objectives. They may also point to areas for improved coordination, either across the federal agencies, with other governments and authorities, or with nongovernmental organizations and the private sector.

Step by step, we will build science-based, intelligence-driven fisheries management with monitoring, enforcement, and real consequences, incorporating innovative technologies and increased collaborative international and private sector information sharing. We will make seafood supply chains transparent and tackle forced labor in the fisheries sector. We recognize that our task will extend beyond the 5-year period of this Strategy, as we will continue building real law enforcement and diplomatic consequences for those operating outside of accepted maritime rules–based order. We are driving toward meaningful progress in isolating those that engage in IUU fishing and creating an environment where IUU fishing fleets and their owners no longer benefit economically. In all of our endeavors, we will foster and strengthen partnerships with other governments and authorities, the nonprofit conservation community, and the private sector to turn the tide of this global scourge. It is only by working together that we can make strides in collective and sound fisheries management and maritime governance.

Appendix A: Priority Regions

Section 3552 (b) of the Act charges the Maritime SAFE Act Interagency Working Group on IUU Fishing with developing a list of priority regions at risk for IUU fishing. The work is intended to help focus and prioritize work through the Department of State's overseas Missions to support capacity building, training, and information sharing to combat IUU fishing, as well as a wide range of activities by all of the Maritime SAFE Act Working Group agencies and the equities they represent.

The Maritime SAFE Act defines "priority regions" as "at high risk for IUU fishing activity or the entry of illegally caught seafood into the markets of countries in the region; and in which countries lack the capacity to fully address the illegal activity."

The Working Group assessed different regions through a framework that evaluated information about recent or egregious cases of IUU fishing in each region, the diverse coastal countries' institutional and operational capacity to deal with IUU fishing, and how the countries within the region participate in the seafood supply chain and global markets.

Based on this analysis, the Working Group developed the following list of priority regions, along with an illustrative list of the countries, territories, and entities with coastlines in or near each region, tiered to help prioritize U.S. activities to address the risks of IUU fishing (Figure A1). Regions are not ranked within each tier.

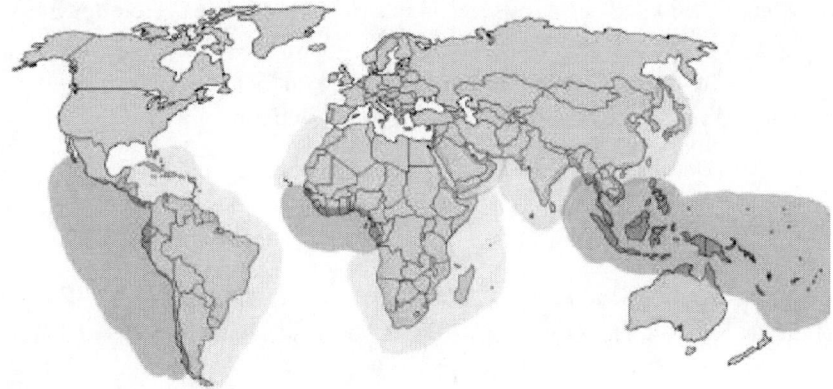

Figure A1. Priority regions categorized into three tiers. Tier 1 priority regions are shown in purple, Tier 2 in yellow, and Tier 3 in blue.

Tier One

These are regions where there was both clear information about the challenges resulting from IUU fishing and ample existing opportunities for U.S. partnerships and activities that could address those challenges.

South and Central America (Pacific Ocean)

Chile, Colombia, Costa Rica, El Salvador, Ecuador, Guatemala, Honduras, Mexico, Panama, Peru.

Gulf of Guinea

Benin, Cameroon, Cabo Verde, Côte d'Ivoire, Democratic Republic of Congo, Equatorial Guinea, Gabon, Ghana, Guinea, Guinea-Bissau, Liberia, Nigeria, Republic of Congo, São Tome & Principe, Sierra Leone, Togo.

Southeast Asia (Gulf of Thailand, Java Sea, Banda Sea, Celebes Sea)

Brunei, Burma, Cambodia, Indonesia, Malaysia, Papua New Guinea, Philippines, Singapore, Thailand, Timor Leste, Vietnam.

Pacific Islands

Australia, Cook Islands, Federated States of Micronesia, Fiji, French Polynesia, Kiribati, Nauru, New Caledonia, New Zealand, Niue, Palau, Papua New Guinea, Pitcairn Islands, Republic of the Marshall Islands, Samoa, Solomon Islands, Tokelau, Tonga, Tuvalu, Vanuatu, Wallis and Futuna.

Tier Two

While we had some specific information and known opportunities for cooperative work in the regions in Tier Two, we could improve our understanding of the situation. U.S. agencies and our partners are looking for opportunities to build law enforcement cooperation, share information, and support training and capacity building within these regions.

Central America and Caribbean (Gulf of Mexico and Caribbean Sea)

Antigua and Barbuda, Bahamas, Barbados, Belize, Colombia, Cuba, Costa Rica, Dominica, Dominican Republic, France (Guadeloupe, Martinique, Saint-Barthélemy, Saint Martin, French Guiana), Grenada, Guatemala, Haiti, Honduras, Jamaica, Kingdom of the Netherlands (Aruba, Bonaire, Curaçao, Saba, Sint Eustatius, and Sint Maarten), Mexico, Nicaragua, Panama, Saint Kitts and Nevis, Saint Lucia, Saint Vincent and the Grenadines, Trinidad and Tobago, United Kingdom (Cayman Islands, British Virgin Islands, Anguilla, Montserrat, Turks and Caicos Islands), Venezuela.

South America (Atlantic Ocean)

Argentina, Brazil, French Guiana, Guyana, Suriname, and Uruguay.

Northwest Africa (Atlantic Ocean)

The Gambia, Mauritania, Morocco, Senegal.

Southern and Central Africa (Atlantic and Indian Ocean)

Angola, Madagascar, Mozambique, Namibia, South Africa.

East Africa (Indian Ocean)

Comoros Islands, Djibouti, Eritrea, Kenya, Madagascar, Mozambique, Mauritius, Seychelles, Somalia, Sudan, Tanzania.

Tier Three

IUU fishing has been raised as a concern in these regions, though details are limited. With this designation, the Working Group aims to get a better understanding of the IUU fishing problem in each of these regions.

Middle East and Gulf States (Persian Gulf, Gulf of Oman, Gulf of Aden, Red Sea)

Bahrain, Egypt, Iran, Iraq, Israel, Jordan, Kuwait, Oman, Palestinian Territories, Qatar, Saudi Arabia, United Arab Emirates, Yemen.

South Asia (Bay of Bengal)

Bangladesh, India, Maldives, Pakistan, Sri Lanka.

East Asia Pacific (East China Sea, Sea of Japan, Sea of Okhotsk)

China (including Hong Kong and Macau), Japan, Democratic People's Republic of Korea (DPRK), Republic of Korea, Russian Federation, Taiwan.

Appendix B: Priority Flag States

Section 3552 (b) of the Maritime Security and Fisheries Enforcement (SAFE) Act charges the Interagency Working Group on Illegal, Unreported, and Unregulated (IUU) Fishing with developing a list of "priority flag states" to be the focus of U.S. assistance. According to the Act, a priority flag state's vessels: "actively engage in, knowingly profit from, or are complicit in IUU fishing" and, at the same time, the priority flag state "is willing, but lacks the capacity, to monitor or take effective enforcement action against its fleet."

The Working Group looked at the countries and authorities within the priority regions it had previously determined and considered several criteria to understand the full picture of the IUU fishing problem within those regions better, including:

- Available information about IUU fishing activities by different fleets;
- Authorities' institutional and operational capacity to monitor and police their fleets and waters effectively; and
- Willingness to work with the United States to remedy IUU fishing activities by their vessels.

Based on this analysis, the Working Group selected five priority flag states and authorities to be the focus of its engagement, assistance, and resources to strengthen efforts against IUU fishing: Ecuador, Panama, Senegal, Taiwan, and Vietnam (Figure B1).

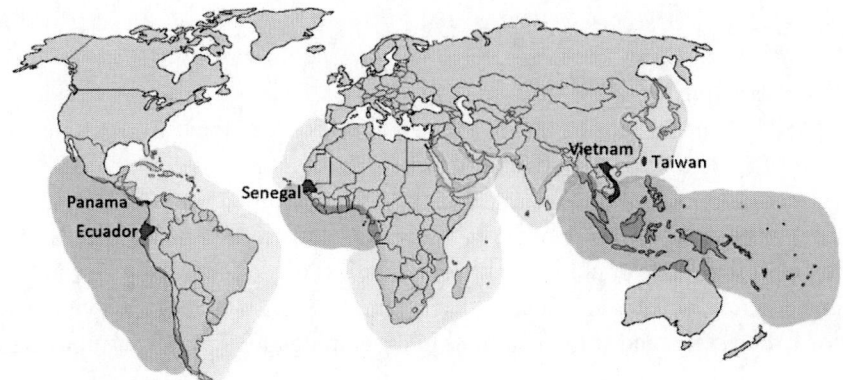

Figure B1. Priority flag states and administrations overlaid with the tiered priority regions.

Each of the five has demonstrated a clear willingness and interest to take effective action against IUU fishing activities associated with their vessels. Their counter-IUU fishing efforts will be supported and enhanced by a positive, cooperative partnership with the United States. The Working Group will identify diplomatic, military, law enforcement, economic, and capacity-building tools and provide technical assistance to counter IUU fishing such as:

- Law enforcement training and coordination
- Implementation of the Port State Measures Agreement
- Capacity building, including for investigations and prosecution, developing strong access agreements, and improving information sharing
- Developing and implementing comprehensive seafood traceability systems
- Promoting and encouraging the mandated use of vessel tracking technologies, such as vessel monitoring systems
- Negotiating, amending, and/or improving implementation of shiprider agreements
- Engaging in other collaborative international fisheries, maritime security, and marine conservation initiatives
- Promoting and supporting efforts to combat labor abuses in the seafood sector

FAQs

Q: What are the next steps?

A: The Department of State will develop resources for U.S. missions in priority flag states and administrations and other countries within the priority regions to support the development of projects and initiatives to combat IUU fishing. Activities will reflect the IUU fishing concerns and needs specific to a region or flag state or administration and will take into account U.S. projects and activities already underway. The United States will engage with the priority flag states and administrations in the coming months to determine how best to provide support.

Q: What is the difference between the priority regions and priority flag states?

A: While both "priority" designations reflect concerns about a lack of capacity to address IUU fishing, priority regions are geographic areas where there is a high risk of IUU fishing occurring, or of illegally caught seafood entering the local and regional markets. Priority flag states or administrations are those that have issues of IUU fishing occurring with their fleets and have demonstrated a willingness to address those issues, but may lack capacity. Priority regions struggle with the problem of IUU fishing from a coastal state and market state perspective, and priority flag states or administrations struggle with IUU fishing by their flagged vessels. All the priority flag states or administrations are also located within a priority region; they will benefit from cooperative activities to prevent and deter IUU fishing in the priority regions' waters.

Q: Why aren't all flag states known to engage in IUU fishing included on this list?

A: The Act's criteria call for more than just determining which flag states have vessels that actively engage in IUU fishing; priority flag states or administrations must have the political will to address the problem and lack the capacity to do so effectively. As such, some nations that are also known to have vessels engaging in IUU fishing are not priority flag states in this context because they have the necessary operational and institutional capacity to prevent and deter these activities, though they are not effectively using it.

Q: Will this list be updated or revised (i.e., how long is the "priority" designation valid)? A: Yes, the list will be updated and revised. The process used to develop this list is an iterative one; the Task Group will continue to update both the components of the framework used to determine the priority flag states or administrations and the list itself.

Appendix C: Working Group Membership

The Maritime Security and Fisheries Enforcement Act (Maritime SAFE Act) was passed in December 2019 to provide a whole-of-government approach to counter illegal, unreported and unregulated (IUU) fishing and related threats to maritime security. Section 3551 of the Act requires the establishment of a Working Group, consisting of 21 relevant U.S. federal agencies, to strengthen maritime security and combat IUU fishing.

The 21 U.S. agencies of the Working Group each carry out different activities to confront and deter impacts and consequences of IUU fishing activities. With today's interconnected fisheries and seafood markets, combating IUU fishing requires a coordinated and comprehensive approach by the U.S. government. The member agencies of the Working Group are: National Oceanic and Atmospheric Administration (current chair until June 2023), U.S. Coast Guard, Department of State, Department of Defense, United States Navy, United States Agency for International Development, United States Fish and Wildlife Service, Department of Justice, Department of the Treasury, U.S. Customs and Border Protection, U.S. Immigration and Customs Enforcement, Federal Trade Commission, Department of Agriculture, Food and Drug Administration, Department of Labor, Director of National Intelligence appointee, National Security Council, Council on Environmental Quality, Office of Management and Budget, Office of Science and Technology Policy, and the Office of the United States Trade Representative.

The Working Group sets up mechanisms for the agencies to share information, pool their expertise and efforts, provide technical assistance, and collectively work with governments, authorities, and the private sector to combat IUU fishing and strengthen maritime security. The Working Group has currently established four subworking groups to assist in carrying out its responsibilities: Gulf of Mexico IUU Fishing; Maritime Intelligence Coordination; Labor in the Seafood Supply Chain, Including Forced Labor; and Public-Private Partnerships. In addition, the Working Group has convened shorter term task groups to address specific tasks such as the identification of priority regions and flag states and development of this Strategy.

Chapter 6

Tackling Illicit Fishing at Sea and Ports Before It Ends Up on Your Plate[*]

United States Government Accountability Office

Attention seafood lovers—some of your favorite imported fish might be vulnerable to illegal, unreported, or unregulated fishing practices. These illicit fishing practices can include a number of activities, such as using prohibited gear or violating international agreements. These activities undermine the economic and environmental sustainability of the fishing industry in the U.S. and globally. Additionally, they can jeopardize food security and contribute to transnational crime by supporting criminal networks—such as narcotics trafficking—or other illegal activities at sea.

In today's WatchBlog post, we'll look at our recent work on how agencies are tackling this issue, both at sea and at U.S. ports, and what more needs to be done.

Combatting Illegal, Unreported, and Unregulated Fishing at Sea

The ocean covers more than 70% of the planet, so identifying illegal, unreported, and unregulated (sometimes known as IUU) fishing can be a difficult task. The U.S. collects and analyzes information from a variety of sources to identify such fishing on the high seas (areas outside of any one

---

[*] This is an edited, reformatted and augmented version of the United States Government Accountability Office Publication, dated June 21, 2023.

In: Ongoing Efforts to Combat Illegal, Unreported ...
Editor: Gordon B. Maddox
ISBN: 979-8-89530-858-5
© 2026 Nova Science Publishers, Inc.

nation's jurisdiction). For example, technology for tracking vessel location at sea helps U.S. agencies identify movements of fishing vessels on the high seas that may indicate illegal, unreported, and unregulated fishing.

Illegal fishing imports.

The U.S. works with other nations through multilateral agreements to collectively prevent illegal, unreported, and unregulated fishing. Agreements may, for example, limit the number and types of fish that can be caught in a specific region. In 2019, officials from Japan, China, Russia, South Korea, Canada, and the U.S. collaborated to patrol areas of the northern Pacific Ocean covered by these types of agreements, and found 58 violations of conservation measures through various multilateral conservation agreements.

Combatting Illegal Fishing Imports

The U.S. imports as much as 70-80% of the seafood we purchase. While it's impossible to pin down how much of that seafood was caught through illegal, unreported, and unregulated fishing, the U.S. International Trade Commission estimated that about 11% of the value of the U.S.'s seafood imports in 2019 were derived from it.

In our new report, we looked at some of the actions federal agencies have taken to prevent seafood caught through these illicit fishing practices from ending up on your plate. For example, the National Marine Fisheries Service

(NMFS) and U.S. Customs and Border Protection (CBP) work together to target such seafood imports. Their efforts include things like monitoring incoming seafood imports that fit a pattern of concern, such as importers with past trade violations. NMFS and CBP also share information in support of this goal. For example, these agencies share information through a CBP interagency group that works to combat a variety of import violations, including importing seafood caught through illegal, unreported, and unregulated fishing. However, NMFS officials told us they weren't always able to get timely information through their collaboration, which could undermine time-sensitive efforts to target, investigate, or identify imports of concern. Because of this, we recommended that CBP work with NMFS to ensure that it has timely access to the information it needs to combat imports of seafood caught through illegal, unreported, and unregulated fishing practices.

Coast Guard Officials Preparing to Board and Inspect a Fishing Vessel.

Chapter 7

Guardians of the Sea: Examining Coast Guard Efforts in Drug Enforcement, Illegal Migration, and IUU Fishing[*]

Caitlin Keating-Bitonti

Chairman Webster, Ranking Member Carbajal, and Members of the Subcommittee, on behalf on the Congressional Research Service (CRS), thank you for this opportunity to appear before you. I am Caitlin Keating-Bitonti, an Analyst in Natural Resources Policy. The Subcommittee requested that CRS testify about the United States Coast Guard's role in the at-sea enforcement of living marine resource laws and international agreements as it pertains to illegal, unreported, and unregulated (IUU) fishing. In accordance with our enabling statutes, CRS takes no position and makes no recommendations on legislative or policy matters. My testimony draws on my own area of specialization at CRS—federal ocean science policy and relevant international agreements—and on the input of other CRS colleagues who cover other issues often associated with ocean policy, including IUU fishing.

Illegal, Unreported, and Unregulated (IUU) Fishing

IUU fishing is an ongoing, multi-faceted global issue that affects the ocean ecosystem and the sustainable management of living marine resources, both

---

[*] This is an edited, reformatted and augmented version of Congressional Research Service Statement Prepared for Committee on Transportation and Infrastructure, Subcommittee on Coast Guard and Maritime Transportation, U.S. House of Representatives, Publication No. TE10089, dated November 14, 2023.

In: Ongoing Efforts to Combat Illegal, Unreported …
Editor: Gordon B. Maddox
ISBN: 979-8-89530-858-5
© 2026 Nova Science Publishers, Inc.

within areas of national jurisdiction and on the high seas.[1] IUU fishing can impact the accuracy of the data needed to inform fisheries conservation and management decisions, thereby adding to overfishing and threatening food security in certain regions. Furthermore, the difficulty in regulating fishing vessels on the high seas may allow some of the vessels involved in IUU fishing to engage in other transnational crimes, such as labor abuses, drug smuggling, and human trafficking.[2]

According to the U.S. Coast Guard, IUU fishing has replaced piracy as the leading global maritime security threat.[3] IUU fishing generally refers to fishing activities that violate national laws or international fisheries conservation and management measures. The international definition of IUU fishing is provided in the Food and Agriculture Organization (FAO) of the United Nations International Plan of Action for IUU fishing.[4] The International Plan of Action for IUU fishing was developed as a voluntary instrument within the framework of the FAO Code of Conduct for Responsible Fisheries, which has the general objective of promoting sustainable fisheries.[5] In general,

- *Illegal fishing* refers to fishing activities conducted in contravention of applicable laws and regulations, including those laws and rules adopted at the regional and international level.
- *Unreported fishing* refers to those fishing activities that are not reported or are misreported to relevant authorities in contravention of national laws and regulations or reporting procedures of a relevant regional fisheries management organization (RFMO). RFMOs are treaty-based international bodies composed of nations that share an

---

[1] National Oceanic and Atmospheric Administration (NOAA), National Marine Fisheries Service (NMFS), *Report to Congress: Improving International Fisheries Management*, August 2023, p. 10.
[2] U.S. Coast Guard (USCG), Illegal, Unreported, and Unregulated Fishing Strategic Outlook, September 2020, p. 2.
[3] Ibid.
[4] Food and Agriculture Organization of the United Nations (FAO), International Plan of Action to Prevent, Deter and Eliminate Illegal, Unreported and Unregulated Fishing, Rome, Italy, 2001, http://www.fao.org/docrep/003/y1224e/y1224e00.HTM. The Agreement on Port State Measures to Prevent, Deter, and Eliminate Illegal, Unreported, and Unregulated Fishing uses the International Plan of Action for Illegal, Unreported, and Unregulated (IUU) Fishing's definition (FAO, Agreement on Port State Measures to Prevent, Deter, and Eliminate Illegal, Unreported and Unregulated Fishing, Rome, Italy, June 20, 2012, ftp://ftp.fao.org/FI/DOCUMENT/PSM/circular_lett_2012.pdf.).
[5] FAO, Code of Conduct for Responsible Fisheries, Rome, Italy, 1995, http://www.fao.org/docrep/005/v9878e/v9878e00.HTM.

- *Unregulated fishing* refers to fishing activities occurring in areas, or fishing for fish stocks,[6] for which there are no applicable conservation and management measures and where such fishing activities are conducted in a manner inconsistent with a nation's or entity's responsibility under international law. *Unregulated fishing* also includes fishing activities conducted by vessels without nationality within the geographic boundaries of an RFMO, or by vessels flying a flag of a nation not a party to the RFMO with authority in that area.[7]

IUU fishing undermines national and regional efforts to conserve and manage fish stocks.[8] FAO estimates that one in five (or 20%) fish caught around the world comes from IUU fishing and, in some regions, such as in West Africa, it can be as high as 40%.[9]

Illegal fishing can entail fishing for nonpermitted species, fishing above management quotas, and fishing out of season. These illegal fishing behaviors can contribute to stocks being fished at biologically unsustainable levels (i.e., at rates greater than species can replenish themselves). FAO estimates that the percentage of stocks fished at biologically unsustainable levels has been increasing since 1970s, from about 10% in 1974 to about 35% in 2019.[10] In particular, in 2019, approximately 77% of catch off the Pacific coast of South America occurred at biologically unsustainable levels.[11]

By its very nature, IUU fishing is difficult to quantify, but there is general global consensus that the impacts of IUU fishing have far-reaching negative

---

[6] NOAA Fisheries defines a *stock* as "a part of a fish population usually with a particular migration pattern, specific spawning grounds, and subject to a distinct fishery. A fish stock may be treated as a total or a spawning stock. Total stock refers to both juveniles and adults, either in numbers or by weight, while spawning stock refers to the numbers or weight of individuals that are old enough to reproduce." NOAA, *NOAA Fisheries Glossary*, p. 49.

[7] NOAA, NMFS, *Report to Congress: Improving International Fisheries Management*, August 2023, p. 10.

[8] FAO, "The Toll of Illegal, Unreported and Unregulated Fishing," at https://www.un.org/en/observances/end-illegal-fishing-day.

[9] FAO, "Four Reasons Illegal, Unreported and Unregulated (IUU) Fishing Affects Us and What We Can Do About It," at https://www.fao.org/fao-stories/article/en/c/1403336/ and NOAA, NMFS, "Understanding Illegal, Unreported, and Unregulated Fishing," at https://www.fisheries.noaa.gov/insight/understanding-illegal-unreported-and-unregulated-fishing.

[10] FAO, 2022 State of World Fisheries and Agriculture, p. 46.

[11] FAO, 2022 State of World Fisheries and Agriculture, p. 47.

consequences.[12] First, IUU fishing undermines the sustainable management of fishery resources—resources that provide both food security and socioeconomic stability in many parts of the world. Developing countries that depend on fisheries for food security and export income are most at risk from IUU fishing.[13] For example, according to a 2022 report by the FAO, aquatic foods provide at least 20% of the average intake of animal protein for 3.3 billion people.[14] IUU fishing can inhibit lawful access to this protein source.

A second negative consequence of IUU fishing is that it provides an unfair advantage to entities thatengage in these activities.[15] For example, vessels conducting IUU fishing avoid operational costs by not complying with regulatory requirements and may earn more revenue by exceeding harvest limits.

Conversely, those fishing legally may be harmed by lower catch rates and higher associated fishing costs. IUU fish in the marketplace can put legal fishers at an economic disadvantage and cause them to lose revenue. According to FAO, IUU fishing catches millions of tons of fish every year,[16] and experts have calculated that IUU costs the global economy up to tens of billions of dollars every year.[17]

Experts note that international cooperation is necessary to manage many fish stocks because some species move among different national zones of jurisdiction and the high seas. However, actions to combat IUU fishing activities are often hindered by the large areas in which fishing takes place, the lack of resources for adequate enforcement, weak governance institutions, and inadequate international cooperation. On the high seas, vessels are subject to the laws of their flag state—the *flag state* of a vessel is the nation of jurisdiction under whose laws the vessel is registered or licensed and is

---

[12] California Environmental Associates, "Distant Water Fishing: Overview of Research Efforts and Current Knowledge," October 2018, p. 7.

[13] NOAA, NMFS, "Understanding Illegal, Unreported, and Unregulated Fishing," at https://www.fisheries.noaa.gov/insight/understanding-illegal-unreported-and-unregulated-fishing.

[14] FAO, *The State of World Fisheries and Aquaculture 2022: Towards Blue Transformation*, Rome, FAO, 2022, pp. 12-13, at https://doi.org/10.4060/cc0461en (hereinafter referred to as FAO, 2022 State of World Fisheries and Agriculture).

[15] NOAA, NMFS, "Understanding Illegal, Unreported, and Unregulated Fishing," at https://www.fisheries.noaa.gov/insight/understanding-illegal-unreported-and-unregulated-fishing.

[16] FAO, "The Toll of Illegal, Unreported and Unregulated Fishing," at https://www.un.org/en/observances/end-illegal-fishing- day.

[17] Enric Sala et al., "The Economics of Fishing the High Seas," *Science Advances*, vol. 4, no. 6 (2018).

deemed the nationality of the vessel.[18] Vessels are also subject to the applicable rules established by international agreements and conventions to which their flag state is a party. The expectation is that all fishing nations exercise responsible flag state control over their vessels, including their distant water fleets operating on the high seas.

China's Role in the Exploitation of Global Fisheries

IUU fishing occurs throughout the world, and according to the U.S. International Trade Commission a portion of the seafood entering the United States reportedly is obtained from IUU fishing activities. The U.S. International Trade Commission estimated that in 2019 about $2.4 billion (or 11%) worth of U.S. seafood imports were products of IUU fishing, of which about $204.3 million were obtained from Chinese IUU fishing.[19]

China is one of the world's largest seafood importers, having imported approximately 4.1 million metric tons of seafood in 2022.[20] Unlike other large importers such as the United States and Japan, the majority of seafood that China imports is not consumed in country.[21] Recent estimates have found that nearly 75% of all fish imported by China never makes it to the Chinese market, but instead is re-exported into the global market.[22]

In recent years, the IUU Fishing Index—a collaboration between Global Initiative Against Transnational Organized Crime, a non-governmental organization, and Poseidon Aquatic Resource Management Ltd., a private fisheries and aquaculture consultancy—has consistently identified China as

---

[18] Article 94 of the United Nations Convention on the Law of the Sea (United Nations, *United Nations Convention on the Law of the Sea of 10 December 1982, Overview and Full Text*, at https://www.un.org/depts/los/convention_agreements/convention_overview_ convention.htm [hereinafter referred to as UNCLOS]). Although the United States is not a party to UNCLOS, some members of the executive branch have stated that some (but not all) portions of UNCLOS reflect *customary international law*.

[19] U.S. International Trade Commission, "Illegal, Unreported, and Unregulated Fishing Accounts for More Than $2 Billion of U.S. Seafood Imports, Reports USITC," press release, March 18, 2021, at https://www.usitc.gov/press_room/news_release/2021/ er0318ll1740.htm.

[20] U.S. Department of Agriculture, Foreign Agriculture Service, "2022 China Fishery Products Annual," February 22, 2023.

[21] Beatrice Crona et al., "China At a Crossroads: An Analysis of China's Changing Seafood Production and Consumption.," *One Earth*, vol. 3, no. 1 (2020), pp. 32-44.

[22] Fangzhou Hu et al., "Development of Fisheries in China," *Reproduction and Breeding*, vol. 1, no. 1 (2021), pp. 64-79.

the worst-performing nation overall in combating IUU fishing.[23] (The IUU Fishing Index analyzes the performances of 152 nations.)[24]

China has the world's largest fishing fleet, with an estimated 564,000 vessels, and in 2020 was the top combined producer of marine and inland water catches, making up nearly 15% of global catches.[25] China also has the largest distant water fishing fleet in the world,[26] with an estimated 2,900 to 3,400 vesselsaccording to the U.S. International Trade Commission.[27] Distant-water fishing is the practice of operating fishing fleets outside of your own nation's Exclusive Economic Zone (EEZ), the zone that extends 200 nautical miles seaward of a coastal nation's shoreline.[28] Distant water fishing fleets operate either on the high seas or foreign EEZs. Overfishing and depleted coastal fish stocks in its national waters have led China's fishing industry to develop a distant-water fishing fleet.[29]

China's distant water fishing fleets are alleged to be increasingly engaging in IUU fishing. A 2022 report by the Environmental Justice Foundation estimates that 95% of Chinese distant water fishing crews have witnessed some form of illegal fishing, including the removal of shark fins and the targeting of endangered and protected marine life.[30] In 2021, the National Oceanic and Atmospheric Administration's (NOAA's) National Marine Fisheries Service (NMFS) issued China a negative certification for IUU fishing, under the authorities of the High Seas Driftnet Fishing Moratorium

---

[23] In 2021, China received a 3.86 score out of 5.00 on the IUU Fishing Index (high scores indicate worse performance). The IUU Index also generally finds that countries with DWF fleets, such as China, have poor scores. IUU Fishing Index, "2021 Country Profile: China," at https://iuufishingindex.net/reports/iuu-fishing-index-country-profile-2021-china.pdf.
[24] Ibid.
[25] FAO, 2022 State of World Fisheries and Agriculture, p. 59.
[26] Raul (Pete) Pedrozo, "China's IUU Fishing Fleet: Pariah of the World's Oceans," *International Law Studies*, vol. 99 (2022), p. 329 (hereinafter referred to as Pedrozo, 2022).
[27] United States International Trade Commission, *Seafood Obtained via Illegal, Unreported, and Unregulated Fishing: U.S. Imports and Economic Impact on U.S. Commercial Fisheries*, February 2021, p. 142. Another report estimates that China's DWF fleet is made up of nearly 17,000 vessels, of which about 12,500 were identified as operating outside internationally recognized China waters between 2017-2018. However, the report cautioned that all of these vessels are not operating currently, simultaneously, or consistently in other countries' or international waters (Overseas Development Institute, "China's Distant- Water Fishing Fleet: Scale, Impact and Governance," June 2020).
[28] Article 56 of UNCLOS gives coastal nations sovereign rights for the purpose of conserving and managing natural resources, including fisheries, among other purposes, in the Exclusive Economic Zone (EEZ).
[29] Pedrozo, 2022, p. 330.
[30] Environmental Justice Foundation, "Global Impact of Illegal Fishing and Human Rights Abuse in China's Vast Distant Water Fleet Revealed," April 5, 2022.

Protection Act (16 U.S.C.§1826j(d)).[31] China denied all allegations made by NMFS.[32]

Like many governments with industrial-scale fishing operations, the Chinese government provides financial and policy support to its fishing industry, including its distant water fishing fleet.[33] This support takes a variety of forms, including fuel subsidies, vessel upgrading/replacement subsidies, and tax incentives. Some analysts argue that some types of distant water fishing would be unprofitable for Chinese vessel operators without government subsidies.[34]

China has adopted some policies to address IUU fishing. However, a 2021 report estimated that at least 183 Chinese distant water fishing vessels, some of which are government-owned or -operated, are involved in IUU fishing, suggesting that the China is not holding its vessels accountable for engaging in IUU activities.[35] Under the U.N. Convention on the Law of the Sea, which China ratified in 1996, the flag state has exclusive jurisdiction over vessels flying its flag on the high seas.[36]

U.S. Government Initiatives Aimed at Combating IUU Fishing

Over the past two decades, successive U.S. administrations and Congresses have recognized that IUU fishing poses a threat to national and regional security and have taken a number of actions to combat IUU fishing broadly. These actions attempt to influence the behavior of foreign fishing fleets through international agreements, organizations, and trade, because most IUU activities occur outside of U.S. jurisdiction.[37]

The United States works with other fishing nations through RFMOs and other multilateral international agreements to sustainably manage high seas fisheries and address IUU fishing globally. Several federal agencies, including

---

[31] NOAA, NMFS, *Improving International Fisheries Management*, Report to Congress, August 2023, pp. 18-19.
[32] Ibid.
[33] Pedrozo, 2022, p. 328.
[34] Ian Urbina, "How China's Expanding Fishing Fleet is Depleting the World's Oceans," August 17, 2020.
[35] IUU Fishing Index, 2021 Report, p. 60.
[36] UNCLOS Article 92.
[37] Actions to combat IUU fishing have included enforcement agreements with partner countries, trade monitoring, implementation and enforcement of international treaties, and broad efforts to promote resource sustainability.

the U.S. Coast Guard, NOAA, the Department of Defense, and the Department of State, engage in various efforts to combat IUU fishing on the high seas and in the EEZs of partner nations. The efforts of these federal agencies include establishing strategic partnerships; improving enforcement tools, such as high seas boarding and inspection; identifying and sharing information about countries that have fishing vessels engaged in IUU fishing activities; and assisting partner nations develop and maintain their own counter IUU fishing capacity, among other lines of effort.[38]

In 2019, the Maritime Security and Fisheries Enforcement Act (Division C, Title XXXV, Subtitle C of P.L. 116-92, 16 U.S.C. §§8001 et seq.), commonly known as the Maritime SAFE Act, passed in the National Defense Authorization Act for Fiscal Year 2020. The Maritime SAFE Act seeks to support a whole-of-government approach to counter IUU fishing, improve data sharing, support efforts to counter IUU fishing in priority regions around the world, increase global transparency and traceability across the seafood chain, improve global enforcement operations against IUU fishing, and prevent the use of IUU fishing as a financing source for transnational crime.[39] The Maritime SAFE Act also established the Interagency Working Group on IUU Fishing to support and coordinate a government-wide effort to address IUU fishing globally. The IWG on IUU Fishing is made up of representatives from 21 federal agencies and is currently chaired by a representative from the Department of State, with representatives from the National Oceanic and Atmospheric Administration and U.S. Coast Guard serving as Deputy Chairs.[40]

U.S. Coast Guard's Role in Addressing IUU Fishing

The U.S. Coast Guard is a multi-mission maritime service with the authority to conduct maritime law enforcement operations, including operations aimed at combating IUU fishing activity.[41] The U.S. Coast Guard enforces U.S. and international living marine resources laws in the U.S. EEZ and in key areas of

---

[38] NOAA, NMFS, *Improving International Fisheries Management*, Report to Congress, August 2023, p. 3.
[39] 16 U.S.C. §8002.
[40] 16 U.S.C. §8031(b). The chair of the Interagency Working Group on IUU Fishing rotates every three years among the Secretary of the Department in which the U.S. Coast Guard is operating (i.e., the Department of Homeland Security), Secretary of State, and NOAA Administrator.
[41] 14 U.S.C §102.

the high seas, and works with NOAA, Department of Defense, and Department of State to provide whole- of-government approach to addressing IUU fishing.[42]

The U.S. Coast Guard is the lead U.S. agency for at-sea enforcement of fishery conservation on the high seas.[43] On the high seas, RFMOs manage and conserve fish stocks of a particular species or group of species within a particular geographic area. The 1995 U.N. Fish Stocks Agreement provides an enhanced framework for RFMOs' conservation and management of transboundary fish stocks.[44] Under the 1995 U.N. Fish Stocks Agreement, party nations are obligated to regulate "the activities of vessels flying their flag which fish for such stocks on the high seas."[45] In addition, the agreement gives party nations the right to monitor and inspect vessels of other nation parties to ensure compliance with internationally agreed fishing regulations, including regulations established by RFMOs. Violations of RFMO conservation measures are generally considered IUU fishing.

Both the Interagency Working Group on IUU Fishing's National 5-Year Strategy for Combating Illegal, Unreported, and Unregulated Fishing and the U.S. Coast Guard's IUU Fishing Strategic Outlook Implementation Plan identify strategies used by the U.S. Coast Guard to counter IUU fishing on the high seas, such as at-sea operations, use of vessel tracking data to identify vessels suspected of IUU fishing, and cooperation in partner nation capacity-building exercises.[46]

---

[42] USCG, Illegal, Unreported, and Unregulated Fishing Strategic Outlook, September 2020, p. 4.
[43] USCG, Illegal, Unreported, and Unregulated Fishing Strategic Outlook, September 2020, p. 19.
[44] United Nations, Agreement for the Implementation of the Provisions of the United Nations Convention on the Law of the Sea of 10 December 1982 Relating to the Conservation and Management of Straddling Fish Stocks and Highly Migratory Fish Stocks, at https://www.un.org/depts/los/convention_agreements/texts/fish_stocks_agreement/CONF 164_37.htm (hereinafter referred to as the 1995 U.N. Fish Stock Agreement).
[45] Article 7 of the 1995 U.N. Fish Stock Agreement.
[46] U.S. Interagency Working Group on IUU Fishing, National 5-Year Strategy for Combating Illegal, Unreported, and Unregulated Fishing: 2022-2026, Report to Congress, October 2022, pp. 1-A3-1, and USCG, Illegal, Unreported, and Unregulated Fishing Strategic Outlook Implementation Plan, July 2021, pp. 1-29.

U.S. Coast Guard At-Sea Operations

The U.S. Coast Guard identifies instances of IUU fishing through its at-sea operations. On the high seas, under the authority of some RFMOs, U.S. Coast Guard law enforcement officials may conduct law enforcement boardings and investigations of fishing vessels suspected of IUU fishing.[47] U.S. Coast Guard law enforcement officials may also randomly board other vessels as a means to deter IUU fishing activity. If a U.S. Coast Guard patrol not directly related to IUU fishing suspects a vessel of IUU fishing, the U.S. Coast Guard may provide relevant information to other U.S. federal agencies (e.g., NMFS) for further investigation. The U.S. Coast Guard patrol may also report the suspect vessel to the relevant RFMO to share information about the vessel with other member states of the RFMO to aid in the tracking of the vessel.

The U.S. Coast Guard reports IUU fishing violations identified through at-sea patrol to RFMOs, which alert the vessels flag state. On the high seas, vessels are subject to the laws of their flag state. The U.S. Coast Guard shares information about the vessels it identifies as having participated in IUU fishing to relevant U.S. federal agencies to inform IUU fishing vessel lists,[48] which may trigger port control measures, among other actions.[49]

Experts consider high seas boarding and inspection of vessels to be effective approaches for fisheries law enforcement and for identifying vessels engaged in IUU fishing.[50] According to the U.S. Government Accountability Office (GAO), from 2016 through 2020, the U.S. Coast Guard boarded and inspected 227 fishing vessels on the high seas within the boundaries of RFMOs to which the United States is a party.[51] During these inspections, the U.S. Coast Guard found 90 potential violations of RFMO fishery conservation

---

[47] According to NOAA, the United States is a member of nine multilateral RFMOs. NOAA, NMFS, "International and Regional Fisheries Management Organizations," at https://www.fisheries.noaa.gov/international-affairs/international-and-regional- fisheries-management-organizations.

[48] Pursuant to its statutory requirements under the High Seas Driftnet Fishing Moratorium Protection Act, NMFS prepares a biennial report to Congress that includes a list of nations whose flagged vessels were identified for IUU fishing (16 U.S.C.§1826h).

[49] NOAA, "Frequent Questions: Implementing the Port State Measures Agreement," at https://www.fisheries.noaa.gov/ enforcement/frequent-questions-implementing-port-state-measures-agreement.

[50] For example, FAO, *High Seas Boarding and Inspection of Fishing Vessels: Discussion of Goals, Comparison of Existing Schemes and Draft Language,* September 2003, pp. 1-41.

[51] U.S. Government Accountability Office (GAO), Combating Illegal Fishing: Clear Authority Could Enhance U.S. Efforts to Partner with Other Nations at Sea, GAO-22-104234, November 2021, p. 19.

and management measures.⁵² The information obtained by the U.S. Coast Guard through vessel boardings and inspections can inform U.S. diplomatic engagements with foreign nations. However, only a subset of RFMOs have high seas boarding and inspection measures. The Interagency Working Group on IUU Fishing identified the need for more RFMOs to adopt high seas boarding and inspection measures.⁵³ According to GAO, the U.S. Coast Guard, Department of State, and NOAA are working to promote the adoption of high seas boarding and inspection measures in all RFMOs to which the U.S. is a member.⁵⁴

U.S. Coast Guard Use of Vessel Tracking Data

The scale of the ocean environment enables some fishing fleets to conduct IUU fishing activity unnoticed and presents law enforcement challenges. The International Maritime Organization and other management bodies require large ships, including many commercial fishing vessels, to broadcast their position with an automatic identification system (AIS).⁵⁵ In addition to broadcasting the location of the vessel, AIS devices also broadcast information about the identity, course and speed of the vessel. Radio stations and satellites pick up this information, making vessels trackable even in the most remote areas of the ocean.

The U.S. Coast Guard analyzes vessel tracking data to identify movement patterns that may be indicative of IUU fishing activity.⁵⁶ Fishing vessels that "go dark" by ceasing to broadcast position information may suggest that these vessels are engaging in IUU fishing activities. Research conducted by NOAA, the University of Santa Cruz, and Global Fishing Watch found that vessels most often go dark while fishing next to EEZs with contested boundaries, fishing in EEZ with limited management oversight, and during the transfer of

---

⁵² Ibid.
⁵³ U.S. Interagency Working Group on IUU Fishing, National 5-Year Strategy for Combating Illegal, Unreported, and Unregulated Fishing: 2022-2026, Report to Congress, October 2022, p. 12.
⁵⁴ GAO, Combating Illegal Fishing: Clear Authority Could Enhance U.S. Efforts to Partner with Other Nations at Sea, GAO-22- 104234, November 2021, p. 19.
⁵⁵ Global Fishing Watch, "What Is AIS?," at https://globalfishingwatch.org/faqs/what-is-ais/.
⁵⁶ GAO, Combating Illegal Fishing: Clear Authority Could Enhance U.S. Efforts to Partner with Other Nations at Sea, GAO-22- 104234, November 2021, p. 17.

fish between fishing vessels and refrigerated cargo vessels.[57] The U.S. Coast Guard analyzes vessel tracking data to help guide at-sea patrol operations to target suspect vessels.

In its IUU Fishing Strategic Outlook Implementation Plan, the U.S. Coast Guard acknowledged that it will continue to advance and implement innovative technology to counter IUU fishing and to expand multilateral fisheries enforcement cooperation with partner nations.[58]

U.S. Coast Guard Efforts to Build Capacity for Partner Nations

The U.S. Coast Guard works with partner nations to develop and maintain their own counter IUU fishing capacity, including the enforcement of their own fisheries conservation measures and the investigation and prosecution of their own fishing fleets suspected of IUU fishing. According to its *IUU Fishing Strategic Outlook Implementation Plan*, the U.S. Coast Guard aims to create regionally based international fisheries law enforcement symposiums for foreign partners, support expanded unclassified information sharing about illicit operations, and add counter-IUU fishing measures to existing bilateral agreements, among other initiatives to combat IUU fishing.[59]

One strategy used by the U.S. Coast Guard to help foreign partners build capacity for counting IUU fishing is through shiprider agreements.[60] Shiprider agreements authorize a law enforcement official of one party to embark on a law enforcement vessel, or aircraft, of the other party and exercise certain authorities. U.S. shiprider agreements are designed to allow U.S. law enforcement officials, typically U.S. Coast Guard law enforcement officials, to assist partner nations in combating various illicit maritime activity, such as IUU fishing. In general, U.S. bilateral shiprider agreements allow maritime

---

[57] NOAA, NMFS, "Learning More about "Dark" Fishing Vessels' Activities at Sea," November, 2, 2022, at https://www.fisheries.noaa.gov/feature-story/learning-more-about-dark-fishing-vessels-activities-sea.

[58] USCG, Illegal, Unreported, and Unregulated Fishing Strategic Outlook Implementation Plan, July 2021, pp. 26-27.

[59] USCG, Illegal, Unreported, and Unregulated Fishing Strategic Outlook Implementation Plan, July 2021, pp. 26-27.

[60] U.S. Interagency Working Group on IUU Fishing, National 5-Year Strategy for Combating Illegal, Unreported, and Unregulated Fishing: 2022-2026, Report to Congress, October 2022, p. 15.

law enforcement officers of a partner nation to embark on vessels (and/or aircraft) of the U.S. government.

The presence of a shiprider on board a U.S. government vessel allows the vessel to enforce the laws and regulations of the partner nation, including the boarding and inspection of suspect vessels, within the partner nation's designated territorial sea or exclusive economic zone. Certain shiprider agreements also allow U.S. government vessels with embarked shipriders to pursue flag ships of the party on the high seas.

Not all U.S. bilateral shiprider agreements include counter-IUU fishing provisions. According to GAO, the United States has entered into 15 shiprider agreements that address IUU fishing.[61] One priority of the Interagency Working Group on IUU Fishing is for the U.S. government to establish new bilateral shiprider agreements that have counter-IUU fishing provisions with countries located within priority regions and to add counter-IUU fishing provisions to existing shiprider agreements.[62]

The U.S. Coast Guard also coordinates with the Department of Defense in their at-sea exercises. Some of these exercises may be designed to help partner nations build maritime security capacity, including their capacity to address IUU fishing in their territorial waters and IUU fishing committed by their flagged vessels. For example, the U.S. Coast Guard and U.S. Africa Command collaborated to enhance partner nation maritime enforcement capabilities to counter IUU fishing and other issues.[63]

In its FY2024 Budget Overview, the U.S. Coast Guard also identified it has operational priorities, including capacity building partnerships, aimed at combatting IUU fishing off the east and west coasts of South America, the west coast of Africa, and across Oceania.[64]

---

[61] GAO, Combating Illegal Fishing: Clear Authority Could Enhance U.S. Efforts to Partner with Other Nations at Sea, GAO-22- 104234, November 2021, p. 13.

[62] The Maritime Security and Fisheries Enforcement Act (P.L. 116-92) directs select federal officials to "exercise existing shiprider agreements and to enter into and implement new shiprider agreements" (16 U.S.C. §8013(b)(2)).

[63] For example, the U.S. Coast Guard and the Department of Defense collaborated with partner African nations to help them build maritime security capacity through the U.S. Africa Commands Africa Maritime Law Enforcement Partnership program. See, GAO, Combating Illegal Fishing: Clear Authority Could Enhance U.S. Efforts to Partner with Other Nations at Sea, GAO- 22-104234, November 2021, p. 14.

[64] USCG, Posture Statement: 2024 Budget Overview, pp.10-11.

Conclusion

This concludes my prepared remarks. Thank you for the opportunity to testify, and I welcome your questions. If additional research and analysis related to this issue would be helpful, my CRS colleagues and I stand ready to assist the committee.

Chapter 8

Maritime SAFE Act Interagency Working Group on IUU Fishing Priority Regions[*]

Maritime SAFE Act Interagency Working Group

Section 3552 (b) of the Act charges the Maritime SAFE Act Interagency Working Group on IUU Fishing with developing a list of priority regions at risk for IUU fishing. The work is intended to help focus and prioritize work through the Department of State's overseas Missions to support capacity building, training, and information-sharing to combat IUU fishing, as well as a wide range of activities by all of the Maritime SAFE Act Working Group agencies and the equities they represent.

The Maritime SAFE Act defines "priority regions" as "at high risk for IUU fishing activity or the entry of illegally caught seafood into the markets of countries in the region; and in which countries lack the capacity to fully address the illegal activity."

The Interagency Working Group assessed different regions through a framework that evaluated information about recent or egregious cases of IUU fishing in each region, the diverse coastal countries' institutional and operational capacity to deal with IUU fishing, and how the countries within the region participate in the seafood supply chain and global markets.

---

[*] This is an edited, reformatted and augmented version of Maritime SAFE Act Interagency Publication.

In: Ongoing Efforts to Combat Illegal, Unreported …
Editor: Gordon B. Maddox
ISBN: 979-8-89530-858-5
© 2026 Nova Science Publishers, Inc.

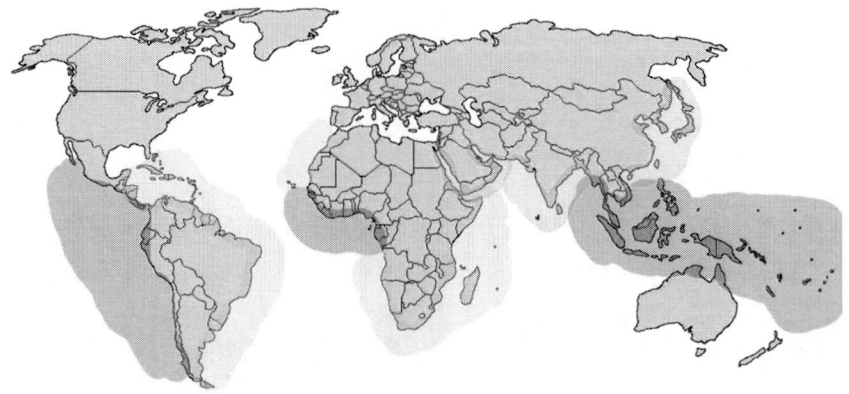

Based on this analysis, the Working Group developed the following list of priority regions, along with an illustrative list of the countries, territories, and entities with coastlines in or near each region, tiered to help prioritize U.S. activities to address the risks of IUU fishing. Regions are not ranked within each tier.

Tier One

South and Central America (Pacific Ocean)

Chile, Colombia, Costa Rica, El Salvador, Ecuador, Guatemala, Honduras, Mexico, Panama, Peru.

Gulf of Guinea

Benin, Cameroon, Cabo Verde, Côte d'Ivoire, Democratic Republic of Congo, Equatorial Guinea, Gabon, Ghana, Guinea, Guinea-Bissau, Liberia, Nigeria, Republic of Congo, São Tome & Principe, Sierra Leone, Togo.

Southeast Asia (Gulf of Thailand, Java Sea, Banda Sea, Celebes Sea)

Brunei, Burma, Cambodia, Indonesia, Malaysia, Papua New Guinea, Philippines, Singapore, Thailand, Timor Leste, Vietnam.

South Pacific

Australia, Cook Islands, Federated States of Micronesia, Fiji, French Polynesia, Kiribati, Nauru, New Caledonia, New Zealand, Niue, Palau, Papua New Guinea, Pitcairn Islands, Republic of the Marshall Islands, Samoa, Solomon Islands, Tokelau, Tonga, Tuvalu, Vanuatu, Wallis and Futuna.

Tier Two

Central America and Caribbean (Gulf of Mexico and Caribbean Sea)

Antigua and Barbuda, Bahamas, Barbados, Belize, Colombia, Cuba, Costa Rica, Dominica, Dominican Republic, France (Guadeloupe, Martinique, Saint-Barthélemy, Saint Martin, French Guiana), Grenada, Guatemala, Haiti, Honduras, Jamaica, Kingdom of the Netherlands (Aruba, Bonaire, Curaçao, Saba, Sint Eustatius, and Sint Maarten), Mexico, Nicaragua, Panama, Saint Kitts and Nevis, Saint Lucia, Saint Vincent and the Grenadines, Trinidad and Tobago, United Kingdom (Cayman Islands, British Virgin Islands, Anguilla, Montserrat, Turks and Caicos Islands), Venezuela.

South America (Atlantic Ocean)

Argentina, Brazil, French Guiana, Guyana, Suriname, and Uruguay.

Northwest Africa (Atlantic Ocean)

The Gambia, Mauritania, Morocco, Senegal.

Southern and Central Africa (Atlantic and Indian Ocean)

Angola, Madagascar, Mozambique, Namibia, South Africa.

East Africa (Indian Ocean)

Comoros Islands, Djibouti, Eritrea, Kenya, Madagascar, Mozambique, Mauritius, Seychelles, Somalia, Sudan, Tanzania.

Tier Three

Middle East and Gulf States (Persian Gulf, Gulf of Oman, Gulf of Aden, Red Sea)

Bahrain, Djibouti, Egypt, Eritrea, Iran, Iraq, Israel, Jordan, Kuwait, Oman, Palestinian Territories, Qatar, Saudi Arabia, Somalia, Sudan, United Arab Emirates, Yemen.

South Asia (Bay of Bengal)

Bangladesh, Burma, India, Maldives, Pakistan, Sri Lanka.

East Asia Pacific (East China Sea, Sea of Japan, Sea of Okhotsk)

China, Hong Kong, Japan, Macau, North Korea, Republic of Korea, Russian Federation, Taiwan.

Index

A

Africa, 17, 30, 39, 57, 68, 73, 84, 140, 153, 175, 179, 180
Alaska, 56, 100
allergens, 21
Angola, 153, 179
Argentina, 134, 153, 179
artificial intelligence, 50, 147

B

Bahrain, 153, 180
Bangladesh, 23, 154, 180
Barbados, 35, 152, 179
biodiversity, 5, 33, 43
border crossing, 114
Brazil, 153, 179
Burma, 23, 152, 178, 180

C

Cambodia, 23, 152, 178
Cameroon, 23, 151, 178
capacity building, 2, 39, 45, 127, 128, 131, 140, 141, 142, 145, 150, 152, 175, 177
Caribbean, 17, 39, 49, 73, 86, 110, 134, 152, 179
Chile, 31, 32, 134, 151, 178
China, 12, 23, 24, 26, 27, 28, 30, 31, 32, 35, 39, 45, 69, 100, 106, 154, 160, 167, 168, 169, 180
coast guard, 6, 36, 43, 44, 48, 52, 55, 58, 59, 61, 62, 68, 69, 72, 73, 74, 75, 76, 77, 78, 79, 80, 81, 82, 84, 85, 86, 87, 90, 125, 135, 142, 147, 157, 161, 163, 164, 170, 171, 172, 173, 174, 175
Colombia, 134, 145, 151, 152, 178, 179
Congo, 151, 178
conservation, 2, 4, 5, 7, 9, 10, 11, 14, 15, 17, 25, 30, 32, 33, 34, 35, 37, 43, 60, 64, 65, 69, 78, 103, 104, 107, 128, 129, 130, 132, 133, 136, 138, 144, 149, 150, 155, 160, 164, 165, 171, 172, 174
Cook Islands, 52, 72, 152, 179
Costa Rica, 134, 151, 152, 178, 179
Côte d'Ivoire, 35, 52, 151, 178
criminal investigations, 46
Cuba, 152, 179
Customs and Border Protection, 18, 42, 91, 97, 98, 100, 101, 102, 106, 147, 157, 161

D

data analysis, 59, 76, 98
data collection, 2, 67, 135, 136, 146, 149
Democratic Republic of Congo, 151, 178
Department of Agriculture, 157
Department of Commerce, 61, 67, 89, 100, 103, 119
Departments of Agriculture, 86, 91
distribution, 89, 101
Dominican Republic, 152, 179
drug smuggling, 164
drug trafficking, 87

E

East Asia, 39, 154, 180
economic losses, 13, 20, 58

Index

ecosystem, 8, 133, 136, 163
Ecuador, 23, 40, 52, 127, 131, 133, 134, 145, 151, 154, 178
Egypt, 153, 180
El Salvador, 151, 178
endangered species, 37, 133
environment, 80, 150, 173
environmental impact, 14
Environmental Protection Agency, 86
environmental sustainability, 58, 60, 97, 100, 159
Equatorial Guinea, 151, 178
equities, 150, 177
Eritrea, 153, 180
European Union, 21, 27, 56, 148
expanded trade, 22
exploitation, 22, 45

F

federal law, 103, 107
Fiji, 23, 52, 72, 152, 179
Fish and Wildlife Service, 43, 46, 91, 108, 116, 126, 157
Food and Drug Administration, 18, 91, 108, 116, 126, 147, 157
food chain, 20
food production, 18
food security, vii, 1, 3, 15, 17, 43, 74, 127, 128, 130, 135, 149, 159, 164, 166
France, 35, 71, 141, 152, 179

G

Gabon, 23, 151, 178
Guatemala, 151, 152, 178, 179
Guinea, 23, 39, 53, 140, 151, 152, 178, 179
Gulf of Mexico, 39, 41, 81, 131, 152, 157, 179
Guyana, 134, 153, 179

H

Haiti, 152, 179
Honduras, 23, 151, 152, 178, 179

Hong Kong, 154, 180
human right, 50, 130

I

Iceland, 27
Immigration and Customs Enforcement, 91, 112, 126, 157
import restrictions, 34, 67, 149
India, 31, 32, 106, 154, 180
Indonesia, 17, 23, 32, 145, 152, 178
international law, 2, 53, 63, 65, 104, 165, 167
international trade, 13, 107
Iran, 153, 180
Iraq, 153, 180
Israel, 153, 180
Italy, 11, 17, 27, 35, 70, 164

J

Japan, 12, 27, 31, 32, 39, 69, 141, 148, 154, 160, 167, 180
Jordan, 28, 153, 180

K

Kenya, 23, 153, 180
Korea, 154, 180
Kuwait, 153, 180

M

machine learning, 50, 56, 147
Malaysia, 152, 178
marine fish, 8, 38, 132, 133, 139
marine species, 76, 109
maritime security, 36, 38, 45, 59, 60, 61, 62, 66, 68, 73, 74, 80, 83, 88, 90, 126, 127, 129, 130, 131, 132, 140, 141, 142, 143, 155, 157, 164, 175
Marshall Islands, 53, 72, 152, 179
Mauritania, 23, 153, 179
Mauritius, 153, 180
Mexico, 35, 41, 81, 151, 152, 178, 179
Morocco, 153, 179

Index

Mozambique, 153, 179, 180

N

Namibia, 35, 153, 179
National Oceanic and Atmospheric Administration (NOAA), 1, 6, 8, 10, 32, 42, 55, 57, 61, 67, 77, 83, 90, 97, 100, 103, 157, 164, 168, 170
natural resources, vii, 1, 8, 63, 130, 168
Netherlands, 152, 179
New Zealand, 29, 71, 152, 179
Nicaragua, 152, 179
Nigeria, 151, 178
North Korea, 23, 180
Norway, 17, 27, 31, 32

P

Pacific, 12, 29, 31, 32, 39, 45, 64, 69, 71, 75, 78, 84, 100, 134, 136, 151, 152, 154, 160, 165, 178, 180
Pakistan, 23, 154, 180
Panama, 40, 52, 127, 131, 134, 147, 151, 152, 154, 178, 179
penalties, 22, 49, 107, 111, 115
permit, 47, 109, 111
Persian Gulf, 39, 153, 180
Peru, 31, 32, 145, 151, 178
Philippines, 23, 29, 141, 145, 152, 178
piracy, 87, 142, 164
producers, vii, 1, 13, 22, 32, 127, 130
profit, 39, 80, 127, 154
project, 43, 46, 79, 134, 135
protected areas, 16
protection, 37, 137

R

radar, 56, 75
ratification, 26, 27
real time, 55, 77
regional cooperation, 143

regulations, vii, 1, 3, 8, 22, 42, 51, 55, 59, 65, 73, 74, 75, 88, 89, 103, 108, 109, 133, 135, 139, 164, 171, 175
regulatory requirements, 15, 166
reliability, 94, 122
resolution, 26, 54, 76, 87
resources, 2, 4, 7, 11, 14, 16, 19, 21, 25, 29, 30, 33, 43, 49, 51, 53, 64, 65, 68, 71, 81, 86, 104, 107, 127, 129, 133, 134, 136, 140, 142, 154, 156, 163, 166, 170
restrictions, 10, 11, 15, 37, 46
revenue, 13, 20, 107, 166
rules, 1, 58, 64, 77, 136, 139, 150, 164, 167
Russia, 27, 28, 31, 32, 35, 69, 160

S

safety, 24, 44, 51, 55, 74, 80
salmon, 31, 100, 105
Samoa, 53, 72, 152, 179
sanctions, 144, 149
SAR, 56
Saudi Arabia, 153, 180
science, 128, 133, 136, 149, 163
seafood, vii, 1, 2, 4, 6, 7, 12, 15, 17, 18, 19, 20, 21, 22, 23, 24, 35, 38, 42, 43, 47, 48, 49, 50, 51, 80, 82, 97, 98, 99, 100, 101, 102, 105, 106, 107, 108, 109, 110, 111, 112, 113, 114, 115, 116, 118, 119, 126, 127, 129, 130, 131, 132, 144, 145, 146, 147, 148, 149, 150, 155, 156, 157, 159, 160, 167, 170, 177
security, vii, 1, 3, 5, 18, 19, 43, 44, 45, 52, 60, 66, 68, 71, 73, 74, 80, 88, 100, 126, 127, 130, 134, 140, 141, 147, 157, 166, 169
security forces, 44, 45
Seychelles, 23, 53, 153, 180
shoreline, 6, 7, 8, 9, 12, 168
shrimp, 48, 105, 114
Sierra Leone, 23, 53, 72, 151, 178
Singapore, 152, 178
slavery, 23, 25
smuggling, 5, 142
Somalia, 153, 180

South Africa, 23, 153, 179
South America, 12, 29, 39, 71, 86, 100, 133, 134, 153, 165, 175, 179
South Asia, 17, 39, 154, 180
South Korea, 12, 23, 27, 31, 69, 160
South Pacific, 29, 31, 32, 64, 79, 179
Southeast Asia, 71, 73, 152, 178
sovereignty, 63, 139
Spain, 12, 35
species, 3, 7, 16, 20, 22, 28, 31, 32, 34, 42, 46, 47, 48, 49, 50, 55, 60, 64, 65, 98, 101, 104, 108, 109, 110, 114, 136, 165, 166, 171
Sri Lanka, 154, 180
stability, 3, 17, 19, 74, 86, 166
statutes, 67, 103, 163
stock, 16, 60, 133, 149, 165
subsidies, 6, 13, 14, 27, 28
Sudan, 153, 180
supply chain, 7, 21, 47, 82, 101, 105, 110, 113, 129, 130, 131, 145, 146, 148, 149, 150, 177
surveillance, 14, 38, 66, 71, 73, 128, 130, 132, 136
sustainability, 4, 19, 20, 21, 33, 71, 135, 136, 169
sustainable development, 70, 128, 149
Switzerland, 18, 20

T

Taiwan, 12, 23, 40, 127, 131, 134, 154, 180
Tanzania, 23, 153, 180
target, 26, 59, 70, 76, 98, 105, 109, 112, 113, 115, 118, 129, 142, 147, 161, 174
territorial, 8, 30, 51, 61, 63, 73, 103, 135, 175
Thailand, 23, 39, 145, 152, 178
threats, 15, 36, 40, 46, 66, 86, 87, 116, 126, 127, 130, 132, 142, 157
Togo, 151, 178
Tonga, 53, 72, 152, 179
trade, 6, 12, 13, 19, 20, 22, 37, 38, 42, 46, 47, 60, 98, 100, 106, 107, 108, 109, 110, 113, 126, 127, 132, 136, 138, 145, 147, 148, 149, 161, 169
trade agreement, 147, 149
trafficking, 5, 6, 12, 22, 23, 24, 25, 37, 40, 46, 51, 60, 65, 67, 82, 87, 142, 145, 146, 148, 159, 164
training, 39, 44, 45, 68, 70, 72, 73, 113, 127, 131, 139, 140, 141, 143, 150, 152, 155, 177
transparency, 6, 21, 35, 37, 45, 76, 130, 135, 138, 145, 146, 170
transport, 20, 24, 26, 101, 103
transshipment, 136, 138
treaties, 53, 61, 169
Trinidad and Tobago, 35, 152, 179
Turks, 152, 179
Tuvalu, 53, 72, 152, 179

U

U.S. assistance, 40, 154
U.S. Department of Agriculture, 18, 125, 147, 167
U.S. economy, 7, 144

V

Vanuatu, 23, 53, 72, 152, 179
Venezuela, 152, 179
vessels, 2, 3, 5, 6, 11, 13, 14, 15, 17, 23, 24, 25, 26, 29, 34, 35, 37, 39, 41, 42, 46, 51, 54, 55, 56, 59, 62, 63, 64, 65, 66, 67, 68, 69, 70, 71, 72, 73, 75, 76, 77, 78, 80, 84, 88, 89, 100, 103, 104, 105, 107, 111, 127, 128, 131, 135, 138, 142, 143, 144, 146, 148, 149, 154, 155, 156, 160, 164, 165, 166, 168, 169, 170, 171, 172, 173, 174, 175
Vietnam, 23, 32, 40, 127, 131, 134, 145, 152, 154, 178

W

wages, 23
war, 73

Washington, 18, 26, 27, 50, 62, 82, 83, 84, 95, 100, 101, 123
water, 8, 10, 11, 12, 45, 46, 63, 133, 135, 137, 167, 168, 169
web, 56, 77, 94, 122
West Africa, 3, 12, 19, 20, 46, 165
White House, 23, 46, 49, 80
wholesale, 101
wildlife, 5, 20, 22, 25, 42, 46, 108, 133
withdrawal, 74
workers, 23, 24, 132
working conditions, 130, 146, 148
working groups, 62, 142
working hours, 24
World Bank, 14
World Trade Organization, 6, 13, 14, 27, 146
worldwide, 19, 56, 64, 70, 76, 84, 141
WTO, 13, 27, 28

Y

Yemen, 153, 180